Common Core Math Practice Grade 5

Complete Content Review Plus 2 Full-length Common Core Math Tests

Elise Baniam - Michael Smith

Common Core Math Practice Grade 5
Published in the United State of America By
The Math Notion
Email: info@Mathnotion.com
Web: WWW.MathNotion.com

Copyright © 2020 by the Math Notion. All rights reserved. No part of this publication may be reproduced, stored in a retrieval system, or transmitted in any form or by any means, electronic, mechanical, photocopying, recording, scanning, or otherwise, except as permitted under Section 107 or 108 of the 1976 United States Copyright Ac, without permission of the author.
All inquiries should be addressed to the Math Notion.

ISBN: 978-1-63620-017-0

About the Author

Elise Baniam has been a math instructor for over a decade now. She graduated in Mathematics. Since 2006, Elise has devoted his time to both teaching and developing exceptional math learning materials. As a Math instructor and test prep expert, Elise has worked with thousands of students. She has used the feedback of her students to develop a unique study program that can be used by students to drastically improve their math score fast and effectively.

- **– SAT Math Workbook**
- **– ACT Math Workbook**
- **– ISEE Math Workbooks**
- **– SSAT Math Workbooks**
- **–many Math Education Workbooks**
- **– and some Mathematics books …**

As an experienced Math teacher, Mrs. Baniam employs a variety of formats to help students achieve their goals: she teaches students in large groups, and she provides training materials and textbooks through her website and through Amazon.

You can contact Elise via email at:
Elise@Mathnotion.com

Common Core Math Practice Grade 5

Get the Targeted Practice You Need to Excel on the Math Section of the Common Core Test Grade 5!

Common Core Math Practice Book Grade 5 is **an excellent investment in your future** and the best solution for students who want to maximize their score and minimize study time. Practice is an essential part of preparing for a test and improving a test taker's chance of success. The best way to practice taking a test is by going through lots of Common Core math questions.

High-quality mathematics instruction ensures that students become problem solvers. We believe all students can develop deep conceptual understanding and procedural fluency in mathematics. In doing so, through this math workbook we help our students grapple with real problems, think mathematically, and create solutions.

Common Core Math Practice Book allows you to:

- Reinforce your strengths and improve your weaknesses
- Practice **2500+ realistic** Common Core math practice questions
- Exercise math problems in a variety of formats that provide intensive practice
- Review and study **Two Full-length Common Core Practice Tests** with detailed explanations

...and much more!

This Comprehensive Common Core Math Practice Book is carefully designed to provide only that **clear and concise information** you need.

WWW.MathNotion.com

… So Much More Online!

✓ FREE Math Lessons

✓ More Math Learning Books!

✓ Mathematics Worksheets

✓ Online Math Tutors

For a PDF Version of This Book

Please Visit WWW.MathNotion.com

Common Core Math Practice Grade 5

Contents

Chapter 1: Place Value and Number Sense 11
    Numbers in Standard Form 12
    Number in Expand Form 13
    Odd or Even 14
    Compare Whole Numbers 15
    Pattern 16
    Round Whole Numbers 17
    Answer key Chapter 1 18

Chapter 2: Whole Number Operations 21
    Order of Operations 22
    Estimate Sums 23
    Estimate Differences 24
    Subtract from Whole Thousands. 25
    Multiplication Whole Number 26
    Long Division by Two Digit 27
    Division with Remainders 27
    Dividing Hundreds 28
    Answer key Chapter 2 29

Chapter 3: Number Theory 31
    Factoring 32
    Prime Factorization 33
    Divisibility Rule 34
    Great Common Factor (GCF) 35
    Least Common Multiple (LCM) 36
    Answer key Chapter 3 37

Chapter 4: Fractions 39
    Adding Fractions – Like Denominator 40
    Adding Fractions – Unlike Denominator 41
    Subtracting Fractions – Like Denominator 42
    Subtracting Fractions – Unlike Denominator 43

WWW.MathNotion.com

Converting Mix Numbers 44
Converting improper Fractions 45
Adding Mix Numbers 46
Subtracting Mix Numbers 47
Simplify Fractions.. 48
Multiplying Fractions.. 49
Multiplying Mixed Number 50
Dividing Fractions 51
Dividing Mixed Number 52
Comparing Fractions 53
Answer key Chapter 4... 54

Chapter 5: Decimal 59
Graph Decimals.. 60
Round Decimals 61
Decimals Addition 62
Decimals Subtraction 63
Decimals Multiplication 64
Decimal Division.. 65
Comparing Decimals 66
Convert Fraction to Decimal 67
Answer key Chapter 5... 68

Chapter 6: Exponent and Radicals.. 71
Positive Exponents 72
Negative Exponents 73
Add and subtract Exponents 74
Exponent multiplication 75
Exponent division 76
Scientific Notation 77
Square Roots 78
Simplify Square Roots 79
Answer key Chapter 6... 80

Chapter 7: Ratio, Proportion and Percent... 83
Proportions... 84
Reduce Ratio 85

Common Core Math Practice Grade 5

Word Problems 86
Percent 87
Convert Fraction to Percent 88
Convert Decimal to Percent 89
Answer key Chapter 7... 90

Chapter 8: Measurement 93
Reference Measurement.. 94
Metric Length Measurement 95
Customary Length Measurement 95
Metric Capacity Measurement 96
Customary Capacity Measurement 96
Metric Weight and Mass Measurement 97
Customary Weight and Mass Measurement 97
Time 98
Answers of Worksheets – Chapter 8 99

Chapter 9: Algebraic Expressions 101
Find a Rule... 102
Variables and Expressions... 103
Translate Phrases 104
Distributive Property 105
Evaluate One Variable Expressions 106
Answer key Chapter 9... 107

Chapter 10: Symmetry and Transformations 109
Line Segments 110
Parallel, Perpendicular and Intersecting Lines 111
Identify Lines of Symmetry 112
Lines of Symmetry 113
Identify Three–Dimensional Figures... 114
Vertices, Edges, and Faces 115
Identify Faces of Three–Dimensional Figures 116
Answers of Worksheets – Chapter 10 117

Chapter 11: Geometry 121
Area and Perimeter of Square 122
Area and Perimeter of Rectangle 123

WWW.MathNotion.com

Area and Perimeter of Triangle...124

Area and Perimeter of Trapezoid 125

Area and Perimeter of Parallelogram 126

Circumference and Area of Circle 127

Perimeter of Polygon 128

Volume of Cubes 129

Volume of Rectangle Prism 130

Answer key Chapter 11...131

Chapter 12: Data and Graphs 133

Mean and Median..134

Mode and Range 135

Stem–And–Leaf Plot 136

Dot plots..137

Probability 138

Bar Graph 139

Answer key Chapter 12...140

Common Core Test Review.. 143

Common Core Practice Test 1 147

Common Core Practice Test 2 161

Answers and Explanations... 175

Answer Key...177

Practice Test 1 179

Practice Test 2 185

Chapter 1:
Place Value and Number Sense

Numbers in Standard Form

Write the number in standard form.

1) 10 million 208 thousand 24

2) 72 million 9 thousand 708

3) 121 million 24 thousand 453

4) 541 million 75 thousand 127

5) 90 billion 15 million 68 thousand 15

6) 12 billion 120 million 5

7) 8 billion 114 million 88 thousand

8) 16 billion 28 thousand 785

9) 75 billion 159 thousand 324

10) 41 billion 3 million 8 thousand 25

11) 16 billion 129 thousand 989

12) 65 billion 220 million 6 thousand 2

13) 785 million 124 thousand 97

14) 33 billion 104 million 11 thousand 57

15) 95 billion 424 million

16) 27 billion 77 million 9 thousand 150

Number in Expand Form

Write the number in expand form.

1) 956: _____.

2) 3,800: _____.

3) 52,457: _____.

4) 60,070: _____.

5) 409,389: _____.

6) 76,805: _____.

7) 745,321: _____.

8) 8,146: _____.

9) 19,037: _____.

10) 52,799: _____.

11) 5,125: _____.

12) 400,544: _____.

13) 600,700: _____.

14) 3,080,000: _____.

Odd or Even

Write odd or even.

1) 19 _____

2) 81 _____

3) 456 _____

4) 852 _____

5) 953 _____

6) 183 _____

7) 987 _____

8) 540 _____

9) 777 _____

10) 544 _____

11) 33 _____

12) 4,458 _____

13) 15,159 _____

14) 9,357 _____

15) 3,000 _____

16) 14 _____

17) 257 _____

18) 660 _____

19) 45,789 _____

20) 15,300 _____

21) 452 _____

22) 49,459 _____

23) 84 _____

24) 7,700 _____

25) 6,451 _____

26) 985 _____

Compare Whole Numbers

Compare, writing <, >, or = between the numbers.

1) 40,420 ☐ 41,004

2) 29,460 ☐ 29,640

3) 78,920 ☐ 87,290

4) 34,570 ☐ 33,750

5) 96,328 ☐ 96,238

6) 85,843 ☐ 85,840

7) 76,584 ☐ 76,854

8) 72,998 ☐ 72,989

9) 37,467 ☐ 37,567

10) 48,878 ☐ 49,878

11) 56,660 ☐ 65,660

12) 73,898 ☐ 69,899

13) 89,990 ☐ 98,110

14) 84,760 ☐ 84,670

15) 26,680 ☐ 26,860

16) 86,440 ☐ 86,440

17) 158,980 ☐ 158,890

18) 201,807 ☐ 201,807

19) 243,240 ☐ 243,420

20) 345,566 ☐ 354,655

21) 187,158 ☐ 196,001

22) 137,983 ☐ 137,895

23) 278,788 ☐ 249,988

24) 194,854 ☐ 194,845

25) 219,390 ☐ 291,110

26) 305,288 ☐ 299,999

27) 317,857 ☐ 371,857

28) 405,710 ☐ 405,170

Pattern

Continue this pattern for four more numbers:

1) 1,400; 1,250; 1,100; 950; _____

2) 3,700; 3,500; 3,300; 3,100; _____

3) 4,200; 3,850; 3,500; 3,150; _____

4) 1,650; 1,530; 1,410; 1,290; _____

5) 2,900; 2,650; 2,400; 2,150; _____

6) 5,000; 4,600; 4,200; 3,800; _____

7) 3,950; 3,700; 3,450; 3,200; _____

8) 2,800; 2,675; 2,550; 2,425; _____

9) 3,850; 3,550; 3,250; 2,950; _____

10) 4,700; 4,500; 4,300; 4,100; _____

11) Write a list of five numbers that follows this pattern: Start at 250 and add 200 each time.

_____

Round Whole Numbers

Round to the place of the underlined digit.

1) 7,46<u>7</u>,589 ≈ _____

2) 54<u>6</u>,125 ≈ _____

3) 9,18<u>7</u>,208 ≈ _____

4) 15,68<u>5</u>,807 ≈ _____

5) 5,45<u>4</u>,676 ≈ _____

6) 3,58<u>8</u>,975 ≈ _____

7) 8,368,<u>5</u>19 ≈ _____

8) 27,754,<u>7</u>69 ≈ _____

9) 42,654,<u>4</u>11 ≈ _____

10) 7, <u>6</u>21,879 ≈ _____

11) 19,78<u>8</u>,987 ≈ _____

12) 4,28<u>6</u>,850 ≈ _____

13) 9,2<u>7</u>3,778 ≈ _____

14) 6,48<u>4</u>,684 ≈ _____

15) 5,157,<u>6</u>28 ≈ _____

16) 8,66<u>7</u>,885 ≈ _____

17) 3,56<u>7</u>,980 ≈ _____

18) 8,369,4<u>3</u>2 ≈ _____

19) 24,256,<u>6</u>80 ≈ _____

20) 5,2<u>2</u>9,758 ≈ _____

21) 6,9<u>8</u>7,422 ≈ _____

22) 4,87<u>7</u>,391 ≈ _____

Answer key Chapter 1

Numbers in Standard Form

1) 10,208,024
2) 72,009,708
3) 121,024,453
4) 541,075,127
9) 75,000,159,324
10) 41,003,008,025
11) 16,000,129,989
12) 65,220,006,002
5) 90,015,068,015
6) 12,120,000,005
7) 8,114,088,000
8) 16,000,028,785
13) 785,124,097
14) 33,104,011,057
15) 95,424,000,000
16) 27,077,009,150

Numbers in Expand Form

1) $(9 \times 100) + (5 \times 10) + 6$
2) $(3 \times 1,000) + (8 \times 100)$
3) $(5 \times 10,000) + (2 \times 1,000) + (4 \times 100) + (5 \times 10) + 7$
4) $(6 \times 10,000) + (7 \times 10) + 0$
5) $(4 \times 100,000) + (9 \times 1,000) + (3 \times 100) + (8 \times 10) + 9$
6) $(7 \times 10,000) + (6 \times 1,000) + (8 \times 100) + 5$
7) $(7 \times 100,000) + (4 \times 10,000) + (5 \times 1,000) + (3 \times 100) + (2 \times 10) + 1$
8) $(8 \times 1,000) + (1 \times 100) + (4 \times 10) + 6$
9) $(1 \times 10,000) + (9 \times 1,000) + (3 \times 10) + 7$
10) $(5 \times 10,000) + (2 \times 1,000) + (7 \times 100) + (9 \times 10) + 9$
11) $(5 \times 1,000) + (1 \times 100) + (2 \times 10) + 5$
12) $(4 \times 100,000) + (5 \times 100) + (4 \times 10) + 4$
13) $(6 \times 100,000) + (7 \times 100)$
14) $(3 \times 1,000,000) + (8 \times 10,000)$

Odd or Even

1) Odd
2) Odd
3) Even
4) Even
5) Odd
6) Odd
7) Odd
8) Even
9) Odd
10) Even
11) Odd
12) Even

Common Core Math Practice Grade 5

13) Odd
14) Odd
15) Even
16) Even
17) Odd
18) Even
19) Odd
20) Even
21) Even
22) Odd
23) Even
24) Even
25) Odd
26) Odd

Compare Whole Numbers

1) <
2) <
3) <
4) >
5) >
6) >
7) <
8) >
9) <
10) <
11) <
12) >
13) <
14) >
15) <
16) =
17) >
18) =
19) <
20) <
21) <
22) >
23) >
24) >
25) <
26) >
27) <
28) >

Pattern

1) 800; 650; 500; 350
2) 2,900; 2,700; 2,500; 2,300
3) 2,800; 2,450; 2,100; 1,750
4) 1,170; 1,050; 930; 810
5) 1,900; 1,650; 1,400; 1,150
6) 3,400; 3,000; 2,600; 2,200
7) 2,950; 2,700; 2,450; 2,200
8) 2,300; 2,175; 2,050; 1,925
9) 2,650; 2,350; 2,050; 1,750
10) 3,900; 3,700; 3,500; 3,300
11) 250; 450; 650; 850; 1,050

Round whole number

1) 7,470,000
2) 546,000
3) 9,187,000
4) 15,686,000
5) 5,455,000
6) 3,589,000
7) 8,368,500
8) 27,754,800
9) 42,654,400
10) 7,600,000
11) 19,789,000
12) 4,287,000
13) 9,270,000
14) 6,485,000
15) 5,157,600
16) 8,668,000
17) 3,568,000
18) 8,369,430
19) 24,257,000
20) 5,230,000
21) 6,990,000
22) 4,877,000

Chapter 2: Whole Number Operations

Order of Operations

Evaluate each expression.

1) $4 + (3 \times 2) =$

2) $19 - (5 \times 3) =$

3) $(12 \times 4) + 11 =$

4) $(19 - 4) - (2 \times 6) =$

5) $45 + (24 \div 6) =$

6) $(15 \times 3) \div 9 =$

7) $(81 \div 9) \times (-2) =$

8) $(6 \times 9) + (28 - 19) =$

9) $70 + (4 \times 5) + 2 =$

10) $(25 \times 2) \div (5 + 5) =$

11) $(-8) + (11 \times 4) + 12 =$

12) $(5 \times 8) - (40 \div 8) =$

13) $(9 \times 4 \div 3) - (15 + 8) =$

14) $(12 + 5 - 10) \times 7 - 10 =$

15) $(48 - 28 + 10) \times (88 \div 11) =$

16) $32 + (18 - (27 \div 9)) =$

17) $(8 + 17 - 3 - 6) + (36 \div 4) =$

18) $(75 - 25) + (18 - 14 + 6) =$

19) $(30 \times 2) + (15 \times 3) - 12 =$

20) $19 + 17 - (24 \times 4) + 13 =$

Estimate Sums

Estimate the sum by rounding each added to the nearest ten.

1) $29 + 8 =$

2) $38 + 37 =$

3) $26 + 19 =$

4) $47 + 28 =$

5) $11 + 39 =$

6) $18 + 33 =$

7) $58 + 21 =$

8) $46 + 68 =$

9) $58 + 78 =$

10) $51 + 72 =$

11) $56 + 29 =$

12) $56 + 13 =$

13) $46 + 49 =$

14) $28 + 75 =$

15) $76 + 93 =$

16) $36 + 49 =$

17) $56 + 72 =$

18) $12 + 47 =$

19) $93 + 27 =$

20) $15 + 83 =$

21) $34 + 74 =$

22) $43 + 73 =$

23) $124 + 282 =$

24) $68 + 64 =$

Estimate Differences

Estimate the difference by rounding each number to the nearest ten.

1) $49 - 12 =$

2) $62 - 52 =$

3) $86 - 57 =$

4) $61 - 43 =$

5) $77 - 58 =$

6) $39 - 17 =$

7) $56 - 29 =$

8) $78 - 61 =$

9) $89 - 46 =$

10) $91 - 69 =$

11) $98 - 63 =$

12) $63 - 17 =$

13) $94 - 58 =$

14) $69 - 48 =$

15) $82 - 52 =$

16) $46 - 22 =$

17) $68 - 51 =$

18) $91 - 66 =$

19) $78 - 67 =$

20) $96 - 38 =$

21) $71 - 64 =$

22) $87 - 31 =$

23) $96 - 28 =$

24) $86 - 43 =$

Subtract from Whole Thousands.

Find the sum or difference.

1) 1,000 − 9 = ___

2) 3,000 − 7 = ___

3) 8,000 − 5 = ___

4) 6,000 − 20 = ___

5) 9,000 − 10 = ___

6) 5,000 − 35 = ___

7) 7,000 − 70 = ___

8) 4,000 − 15 = ___

9) 3,000 − 9 = ___

10) 6,000 − 40 = ___

11) 9,000 − 300 = ___

12) 8,000 − 6 = ___

13) 7,000 − 25 = ___

14) 5,000 − 150 = ___

15) 8,000 − 450 = ___

16) 4,000 − 40 = ___

17) 6,000 − 110 = ___

18) 9,000 − 150 = ___

19) 5,000 − 600 = ___

20) 3,000 − 18 = ___

21) 7,000 − 140 = ___

22) 3,000 − 55 = ___

23) 2,000 − 50 = ___

24) 4,000 − 12 = ___

25) 6,000 − 35 = ___

26) 9,000 − 80 = ___

27) 5,000 − 40 = ___

28) 3,000 − 90 = ___

Multiplication Whole Number

Calculate.

1) $170 \times 8 =$

2) $145 \times 30 =$

3) $(-9) \times 7 \times (-6) =$

4) $-6 \times (-8) \times (-8) =$

5) $14 \times (-14) =$

6) $50 \times (-6) =$

7) $9 \times (-3) \times 7 =$

8) $(-300) \times (-20) =$

9) $(-40) \times (-40) \times 2 =$

10) $150 \times 4 =$

11) $135 \times 30 =$

12) $312 \div 12 =$

13) $(-3,300) \div 4 =$

14) $(-36) \div (-9) =$

15) $272 \div (-17) =$

16) $3,000 \div 24 =$

17) $(-168) \div 4 =$

18) $3,096 \div 3 =$

19) $1,044 \div (-29) =$

20) $5,880 \div 7 =$

21) $(-81) \div 3 =$

22) $(-4,000) \div (-40) =$

23) $0 \div 450 =$

24) $(-1,360) \div 8 =$

25) $6,408 \div 712 =$

26) $(-3,150) \div 5 =$

27) $7,268 \div 2 =$

28) $(-108) \div (-9) =$

Long Division by Two Digit

Find the quotient.

1) $14\overline{)448}$

2) $13\overline{)884}$

3) $21\overline{)588}$

4) $22\overline{)286}$

5) $42\overline{)714}$

6) $28\overline{)252}$

7) $19\overline{)589}$

8) $37\overline{)1,887}$

9) $55\overline{)1,100}$

10) $38\overline{)1,368}$

11) $56\overline{)2,632}$

12) $60\overline{)2,880}$

13) $32\overline{)2,912}$

14) $58\overline{)6,322}$

15) $87\overline{)3,828}$

16) $82\overline{)6,642}$

17) $60\overline{)11,040}$

18) $15\overline{)9,240}$

Division with Remainders

Find the quotient with remainder.

1) $14\overline{)715}$

2) $16\overline{)2,750}$

3) $19\overline{)3,243}$

4) $79\overline{)3,478}$

5) $29\overline{)5,340}$

6) $64\overline{)6,843}$

7) $46\overline{)5,861}$

8) $78\overline{)10,340}$

9) $54\overline{)8,802}$

10) $91\overline{)13,018}$

11) $43\overline{)11,419}$

12) $55\overline{)33,872}$

13) $81\overline{)6,538}$

14) $89\overline{)35,092}$

Dividing Hundreds

Find answers.

1) $3,000 \div 300$

2) $3,200 \div 40$

3) $1,000 \div 200$

4) $3,600 \div 900$

5) $4,200 \div 700$

6) $1,200 \div 400$

7) $2,700 \div 900$

8) $3,500 \div 700$

9) $10,200 \div 300$

10) $20,000 \div 400$

11) $5,400 \div 200$

12) $9,600 \div 600$

13) $1,800 \div 300$

14) $7,700 \div 700$

15) $9,600 \div 800$

16) $4,500 \div 500$

17) $21,000 \div 700$

18) $6,600 \div 200$

19) $8,100 \div 100$

20) $15,000 \div 200$

21) $18,000 \div 6,000$

22) $28,000 \div 70$

23) $490 \div 70$

24) $640 \div 80$

Answer key Chapter 2

Order of Operations

1) 10	6) 5	11) 48	16) 47
2) 4	7) −18	12) 35	17) 25
3) 59	8) 63	13) −11	18) 60
4) 3	9) 92	14) 39	19) 93
5) 49	10) 5	15) 240	20) −47

Estimate sums

1) 40	7) 80	13) 100	19) 120
2) 80	8) 120	14) 110	20) 100
3) 50	9) 140	15) 170	21) 100
4) 80	10) 120	16) 90	22) 110
5) 50	11) 90	17) 130	23) 400
6) 50	12) 70	18) 60	24) 130

Estimate differences

1) 40	7) 30	13) 30	19) 10
2) 10	8) 20	14) 20	20) 60
3) 30	9) 40	15) 30	21) 10
4) 20	10) 20	16) 30	22) 60
5) 20	11) 40	17) 20	23) 70
6) 20	12) 40	18) 20	24) 50

Subtract from Whole Thousands

1) 991	10) 5,960	19) 4,400
2) 2,993	11) 8,700	20) 2,982
3) 7,995	12) 7,994	21) 6,860
4) 5,980	13) 6,975	22) 2,945
5) 8,990	14) 4,850	23) 1,950
6) 4,965	15) 7,550	24) 3,988
7) 6,930	16) 3,960	25) 5,965
8) 3,985	17) 5,890	26) 8,920
9) 2,991	18) 8,850	27) 4,960

28) 2,910

Multiplication Whole Number

1) 1,360
2) 4,350
3) 378
4) −576
5) −196
6) −300
7) −189
8) 6,000
9) 3,200
10) 600
11) 4,050
12) 26
13) −825
14) 4
15) −16
16) 125
17) −42
18) 1,032
19) −36
20) 840
21) −27
22) 100
23) 0
24) −170
25) 9
26) −630
27) 3,634
28) 12

Long Division by Two Digit

1) 32
2) 68
3) 28
4) 13
5) 17
6) 9
7) 31
8) 51
9) 20
10) 36
11) 47
12) 48
13) 91
14) 109
15) 44
16) 81
17) 184
18) 616

Division with Remainders

1) 51 R1
2) 171 R14
3) 170 R13
4) 44 R2
5) 184 R4
6) 106 R59
7) 127 R19
8) 132 R44
9) 163 R0
10) 143 R5
11) 265 R24
12) 615 R47
13) 80 R58
14) 394 R26

Dividing Hundreds

1) 10
2) 80
3) 5
4) 4
5) 6
6) 3
7) 3
8) 5
9) 34
10) 50
11) 27
12) 16
13) 6
14) 11
15) 12
16) 9
17) 30
18) 33
19) 81
20) 75
21) 3
22) 400
23) 7
24) 8

Chapter 3: Number Theory

Factoring

Factor, write prime if prime.

1) 15

2) 72

3) 25

4) 42

5) 32

6) 66

7) 34

8) 20

9) 50

10) 35

11) 40

12) 30

13) 49

14) 54

15) 96

16) 108

17) 76

18) 90

19) 100

20) 85

21) 63

22) 24

23) 48

24) 115

25) 51

26) 93

27) 52

28) 105

Prime Factorization

Factor the following numbers to their prime factors.

1.
   16
   / \
   ___

2.
   38
   / \
   ___

3.
   51
   / \
   ___

4.
   12
   / \
   ___

5.
   18
   / \
   ___

6.
   23
   / \
   ___

7.
   46
   / \
   ___

8.
   58
   / \
   ___

9.
   64
   / \
   ___

10.
   82
   / \
   ___

11.
   87
   / \
   ___

12.
   98
   / \
   ___

Divisibility Rule

Apply the divisibility rules to find the factors of each number.

1) 20   2, 3, 4, 5, 6, 9, 10   13) 24   2, 3, 4, 5, 6, 9, 10

2) 84   2, 3, 4, 5, 6, 9, 10   14) 395   2, 3, 4, 5, 6, 9, 10

3) 252   2, 3, 4, 5, 6, 9, 10   15) 920   2, 3, 4, 5, 6, 9, 10

4) 64   2, 3, 4, 5, 6, 9, 10   16) 137   2, 3, 4, 5, 6, 9, 10

5) 220   2, 3, 4, 5, 6, 9, 10   17) 440   2, 3, 4, 5, 6, 9, 10

6) 465   2, 3, 4, 5, 6, 9, 10   18) 360   2, 3, 4, 5, 6, 9, 10

7) 75   2, 3, 4, 5, 6, 9, 10   19) 495   2, 3, 4, 5, 6, 9, 10

8) 120   2, 3, 4, 5, 6, 9, 10   20) 4,870   2, 3, 4, 5, 6, 9, 10

9) 1,125   2, 3, 4, 5, 6, 9, 10   21) 590   2, 3, 4, 5, 6, 9, 10

10) 88   2, 3, 4, 5, 6, 9, 10   22) 326   2, 3, 4, 5, 6, 9, 10

11) 454   2, 3, 4, 5, 6, 9, 10   23) 114   2, 3, 4, 5, 6, 9, 10

12) 155   2, 3, 4, 5, 6, 9, 10   24) 470   2, 3, 4, 5, 6, 9, 10

Great Common Factor (GCF)

Find the GCF of the numbers.

1) 8, 26

2) 16, 44

3) 28, 38

4) 10, 35

5) 18, 48

6) 36, 52

7) 40, 75

8) 50, 45

9) 52, 8

10) 55, 85

11) 74, 94

12) 65, 20

13) 90, 10

14) 12, 34

15) 58, 86

16) 40, 95

17) 14, 56

18) 70, 100, 30

19) 64, 108

20) 63, 91

21) 20, 15, 35

22) 6, 12, 36

23) 25, 35, 70

24) 41, 39

Least Common Multiple (LCM)

Find the LCM of each.

1) 6, 14

2) 12, 18

3) 4, 16, 12

4) 10, 8

5) 10, 2, 15

6) 35, 7

7) 14, 35, 21

8) 8, 7

9) 11, 22, 44

10) 42, 21

11) 24, 72

12) 100, 25

13) 10, 5, 20

14) 15, 60

15) 20, 4, 3

16) 21, 14

17) 34, 17

18) 16, 64

19) 20, 70

20) 9, 39

21) 13, 8

22) 7, 20

23) 45, 63

24) 27, 4

Answer key Chapter 3

Factoring

1) 1, 3, 5, 15
2) 1, 2, 3, 4, 6, 8, 9, 12, 18, 24, 36, 72
3) 1, 5, 25
4) 1, 2, 3, 6, 7, 14, 21, 42
5) 1, 2, 4, 8, 16, 32
6) 1, 2, 3, 6, 11, 22, 33, 66
7) 1, 2, 17, 34
8) 1, 2, 4, 5, 10, 20
9) 1, 2, 5, 10, 25, 50
10) 1, 5, 7, 35
11) 1, 2, 4, 5, 8, 10, 20, 40
12) 1, 2, 3, 5, 6, 10, 15, 30
13) 1, 7, 49
14) 1, 2, 3, 6, 9, 18, 27, 54
15) 1, 2, 3, 4, 6, 8, 12, 16, 24, 32, 48, 96
16) 1, 2, 3, 4, 6, 9, 12, 18, 27, 36, 54, 108
17) 1, 2, 4, 19, 38, 76
18) 1, 2, 3, 5, 6, 9, 10, 15, 18, 30, 45, 90
19) 1, 2, 4, 5, 10, 20, 25, 50, 100
20) 1, 5, 17, 85
21) 1, 3, 7, 9, 21, 63
22) 1, 2, 3, 4, 6, 8, 12, 24
23) 1, 2, 3, 4, 6, 8, 12, 16, 24, 48
24) 1, 5, 23, 115
25) 1, 3, 17, 51
26) 1, 3, 31, 93
27) 1, 2, 4, 13, 26, 52
28) 1, 3, 5, 7, 15, 21, 35, 105

Prime Factorization

1) $2 \times 2 \times 2 \times 2$
2) 2×19
3) 3×17
4) $2 \times 2 \times 3$
5) $2 \times 3 \times 3$
6) 23 is a prime number
7) 2×23
8) 2×29
9) $2 \times 2 \times 2 \times 2 \times 2 \times 2$
10) 2×41
11) 3×29
12) $2 \times 7 \times 7$

Divisibility Rule

1) 20   <u>2</u>, 3, <u>4</u>, <u>5</u>, 6, 9, 10
2) 84   <u>2</u>, <u>3</u>, <u>4</u>, 5, <u>6</u>, 9, 10
3) 252   <u>2</u>, <u>3</u>, <u>4</u>, 5, <u>6</u>, <u>9</u>, 10
4) 64   <u>2</u>, 3, <u>4</u>, 5, 6, 9, 10
5) 220   <u>2</u>, 3, <u>4</u>, <u>5</u>, 6, 9, <u>10</u>
6) 465   2, <u>3</u>, 4, <u>5</u>, 6, 9, 10
7) 75   2, <u>3</u>, 4, <u>5</u>, 6, 9, 10
8) 120   <u>2</u>, <u>3</u>, <u>4</u>, <u>5</u>, <u>6</u>, 9, <u>10</u>
9) 1,125   2, <u>3</u>, 4, <u>5</u>, 6, <u>9</u>, 10
10) 88   <u>2</u>, 3, <u>4</u>, 5, 6, 9, 10
11) 454   <u>2</u>, 3, 4, 5, 6, 9, 10
12) 155   2, 3, 4, <u>5</u>, 6, 9, 10

Common Core Math Practice Grade 5

13) 24   <u>2</u>, <u>3</u>, <u>4</u>, 5, <u>6</u>, 9, 10
14) 395   2, 3, 4, <u>5</u>, 6, 9, 10
15) 920   <u>2</u>, 3, <u>4</u>, <u>5</u>, 6, 9, <u>10</u>
16) 137   2, 3, 4, 5, 6, 9, 10
17) 440   <u>2</u>, 3, <u>4</u>, <u>5</u>, 6, 9, <u>10</u>
18) 360   <u>2</u>, <u>3</u>, <u>4</u>, <u>5</u>, <u>6</u>, <u>9</u>, <u>10</u>

19) 495   2, <u>3</u>, 4, <u>5</u>, 6, <u>9</u>, 10
20) 4,870 <u>2</u>, 3, 4, <u>5</u>, 6, 9, <u>10</u>
21) 590   <u>2</u>, 3, 4, <u>5</u>, 6, 9, <u>10</u>
22) 326   <u>2</u>, 3, 4, 5, 6, 9, 10
23) 114   <u>2</u>, <u>3</u>, 4, 5, <u>6</u>, 9, 10
24) 470   <u>2</u>, 3, 4, <u>5</u>, 6, 9, <u>10</u>

Great Common Factor (GCF)

1) 2
2) 4
3) 2
4) 5
5) 6
6) 4
7) 5
8) 5
9) 4
10) 5
11) 2
12) 5
13) 10
14) 2
15) 2
16) 5
17) 14
18) 10
19) 4
20) 7
21) 5
22) 6
23) 5
24) 1

Least Common Multiple (LCM)

1) 42
2) 36
3) 48
4) 40
5) 30
6) 35
7) 210
8) 56
9) 44
10) 42
11) 72
12) 100
13) 20
14) 60
15) 60
16) 42
17) 34
18) 64
19) 140
20) 117
21) 104
22) 140
23) 315
24) 108

Chapter 4:
Fractions

Adding Fractions – Like Denominator

Find each sum.

1) $\frac{1}{3} + \frac{1}{3} =$

2) $\frac{3}{7} + \frac{1}{7} =$

3) $\frac{2}{9} + \frac{5}{9} =$

4) $\frac{7}{15} + \frac{1}{15} =$

5) $\frac{5}{23} + \frac{4}{23} =$

6) $\frac{8}{29} + \frac{7}{29} =$

7) $\frac{7}{19} + \frac{1}{19} =$

8) $\frac{5}{16} + \frac{1}{16} =$

9) $\frac{5}{31} + \frac{8}{31} =$

10) $\frac{5}{51} + \frac{8}{51} =$

11) $\frac{1}{17} + \frac{3}{17} =$

12) $\frac{2}{13} + \frac{4}{13} =$

13) $\frac{5}{41} + \frac{19}{41} =$

14) $\frac{2}{55} + \frac{9}{55} =$

15) $\frac{5}{21} + \frac{8}{21} =$

16) $\frac{10}{33} + \frac{2}{33} =$

17) $\frac{3}{11} + \frac{3}{11} =$

18) $\frac{24}{67} + \frac{1}{67} =$

19) $\frac{3}{19} + \frac{6}{19} =$

20) $\frac{20}{47} + \frac{13}{47} =$

Adding Fractions – Unlike Denominator

Add the fractions and simplify the answers.

1) $\frac{1}{2} + \frac{1}{6} =$

2) $\frac{2}{3} + \frac{3}{4} =$

3) $\frac{3}{5} + \frac{1}{2} =$

4) $\frac{7}{10} + \frac{1}{3} =$

5) $\frac{4}{15} + \frac{1}{5} =$

6) $\frac{3}{14} + \frac{2}{7} =$

7) $\frac{2}{7} + \frac{1}{3} =$

8) $\frac{1}{20} + \frac{1}{5} =$

9) $\frac{7}{12} + \frac{1}{6} =$

10) $\frac{1}{8} + \frac{3}{4} =$

11) $\frac{4}{21} + \frac{1}{3} =$

12) $\frac{5}{36} + \frac{2}{9} =$

13) $\frac{4}{35} + \frac{3}{7} =$

14) $\frac{2}{33} + \frac{1}{11} =$

15) $\frac{10}{27} + \frac{1}{3} =$

16) $\frac{7}{45} + \frac{4}{9} =$

17)

18) $\frac{1}{7} + \frac{2}{5} =$

19) $\frac{1}{2} + \frac{5}{22} =$

20) $\frac{3}{16} + \frac{1}{4} =$

21) $\frac{5}{16} + \frac{1}{24} =$

22) $\frac{2}{15} + \frac{3}{10} =$

23) $\frac{5}{42} + \frac{4}{21} =$

24) $\frac{1}{36} + \frac{5}{24} =$

Subtracting Fractions – Like Denominator

Find the difference.

1) $\dfrac{6}{5} - \dfrac{1}{5} =$

2) $\dfrac{7}{12} - \dfrac{4}{12} =$

3) $\dfrac{8}{13} - \dfrac{5}{13} =$

4) $\dfrac{19}{6} - \dfrac{7}{6} =$

5) $\dfrac{11}{21} - \dfrac{9}{21} =$

6) $\dfrac{15}{43} - \dfrac{7}{43} =$

7) $\dfrac{9}{23} - \dfrac{3}{23} =$

8) $\dfrac{18}{47} - \dfrac{15}{47} =$

9) $\dfrac{8}{20} - \dfrac{4}{20} =$

10) $\dfrac{34}{48} - \dfrac{17}{48} =$

11) $\dfrac{6}{7} - \dfrac{2}{7} =$

12) $\dfrac{36}{51} - \dfrac{28}{51} =$

13) $\dfrac{9}{11} - \dfrac{5}{11} =$

14) $\dfrac{29}{49} - \dfrac{14}{49} =$

15) $\dfrac{12}{17} - \dfrac{6}{17} =$

16) $\dfrac{17}{23} - \dfrac{11}{23} =$

17) $\dfrac{5}{7} - \dfrac{1}{7} =$

18) $\dfrac{12}{31} - \dfrac{8}{31} =$

19) $\dfrac{38}{61} - \dfrac{29}{61} =$

20) $\dfrac{31}{52} - \dfrac{20}{52} =$

21) $\dfrac{51}{63} - \dfrac{46}{63} =$

22) $\dfrac{55}{83} - \dfrac{25}{83} =$

23) $\dfrac{54}{77} - \dfrac{30}{77} =$

24) $\dfrac{49}{55} - \dfrac{37}{55} =$

Subtracting Fractions – Unlike Denominator

Solve each problem.

1) $\frac{1}{3} - \frac{1}{6} =$

2) $\frac{3}{5} - \frac{1}{9} =$

3) $\frac{1}{4} - \frac{1}{6} =$

4) $\frac{7}{9} - \frac{3}{10} =$

5) $\frac{11}{12} - \frac{5}{24} =$

6) $\frac{9}{16} - \frac{3}{20} =$

7) $\frac{19}{30} - \frac{1}{6} =$

8) $\frac{1}{2} - \frac{7}{15} =$

9) $\frac{3}{5} - \frac{2}{7} =$

10) $\frac{8}{9} - \frac{4}{11} =$

11) $\frac{6}{8} - \frac{7}{48} =$

12) $\frac{3}{4} - \frac{7}{10} =$

13) $\frac{4}{5} - \frac{8}{45} =$

14) $\frac{6}{7} - \frac{3}{10} =$

15) $\frac{11}{12} - \frac{13}{24} =$

16) $\frac{4}{9} - \frac{15}{63} =$

17) $\frac{5}{12} - \frac{5}{16} =$

18) $\frac{5}{8} - \frac{3}{10} =$

19) $\frac{5}{7} - \frac{3}{8} =$

20) $\frac{3}{4} - \frac{21}{44} =$

Converting Mix Numbers

Convert the following mixed numbers into improper fractions.

1) $2\frac{3}{5} =$

2) $3\frac{4}{5} =$

3) $5\frac{3}{4} =$

4) $3\frac{5}{8} =$

5) $6\frac{2}{5} =$

6) $7\frac{8}{9} =$

7) $2\frac{2}{15} =$

8) $3\frac{1}{13} =$

9) $2\frac{1}{12} =$

10) $5\frac{5}{6} =$

11) $7\frac{3}{5} =$

12) $3\frac{9}{10} =$

13) $6\frac{1}{3} =$

14) $5\frac{5}{6} =$

15) $8\frac{2}{5} =$

16) $4\frac{3}{8} =$

17) $3\frac{5}{9} =$

18) $2\frac{3}{14} =$

19) $7\frac{5}{6} =$

20) $5\frac{6}{7} =$

21) $4\frac{7}{8} =$

22) $3\frac{2}{9} =$

23) $2\frac{9}{11} =$

24) $10\frac{5}{7} =$

WWW.MathNotion.com

Common Core Math Practice Grade 5

Converting improper Fractions

Convert the following improper fractions into mixed numbers

1) $\frac{55}{16} =$

2) $\frac{95}{34} =$

3) $\frac{39}{19} =$

4) $\frac{11}{3} =$

5) $\frac{61}{17} =$

6) $\frac{152}{41} =$

7) $\frac{125}{31} =$

8) $\frac{46}{7} =$

9) $\frac{43}{11} =$

10) $\frac{14}{3} =$

11) $\frac{29}{6} =$

12) $\frac{61}{15} =$

13) $\frac{56}{24} =$

14) $\frac{21}{9} =$

15) $\frac{115}{16} =$

16) $\frac{59}{6} =$

17) $\frac{144}{10} =$

18) $\frac{51}{13} =$

19) $\frac{36}{8} =$

20) $\frac{58}{6} =$

21) $\frac{9}{7} =$

22) $\frac{89}{11} =$

23) $\frac{102}{9} =$

24) $\frac{180}{19} =$

WWW.MathNotion.com

Adding Mix Numbers

Add the following fractions.

1) $3\frac{4}{9} + 2\frac{2}{9} =$

2) $3\frac{2}{5} + 2\frac{1}{5} =$

3) $1\frac{1}{7} + 2\frac{3}{7} =$

4) $4\frac{5}{6} + 2\frac{1}{3} =$

5) $1\frac{4}{15} + 2\frac{2}{5} =$

6) $4\frac{1}{3} + 1\frac{3}{4} =$

7) $3\frac{7}{9} + 3\frac{1}{6} =$

8) $3\frac{5}{8} + 2\frac{1}{2} =$

9) $3\frac{3}{4} + 2\frac{1}{4} =$

10) $1\frac{4}{11} + 2\frac{3}{11} =$

11) $4\frac{1}{2} + 2\frac{2}{5} =$

12) $5\frac{1}{4} + 2\frac{5}{6} =$

13) $6\frac{2}{7} + 2\frac{5}{7} =$

14) $3\frac{7}{8} + 2\frac{3}{16} =$

15) $3\frac{3}{4} + 3\frac{2}{9} =$

16) $4\frac{3}{5} + 2\frac{1}{6} =$

17) $7\frac{1}{2} + 6\frac{3}{5} =$

18) $6\frac{2}{3} + 1\frac{5}{12} =$

19) $2\frac{1}{9} + 7\frac{2}{3} =$

20) $4\frac{1}{6} + 2\frac{5}{9} =$

21) $5\frac{2}{3} + 6\frac{3}{4} =$

22) $7\frac{1}{8} + 1\frac{7}{24} =$

23) $5\frac{3}{7} + 4\frac{1}{8} =$

24) $8\frac{1}{3} + 4\frac{3}{4} =$

Subtracting Mix Numbers

Subtract the following fractions.

1) $8\frac{1}{4} - 7\frac{1}{4} =$

2) $5\frac{5}{6} - 5\frac{1}{6} =$

3) $9\frac{7}{11} - 8\frac{3}{11} =$

4) $5\frac{1}{2} - 2\frac{1}{6} =$

5) $4\frac{1}{4} - 1\frac{1}{8} =$

6) $9\frac{1}{3} - 5\frac{3}{7} =$

7) $5\frac{7}{9} - 2\frac{2}{9} =$

8) $8\frac{15}{17} - 5\frac{11}{17} =$

9) $9\frac{11}{14} - 3\frac{5}{14} =$

10) $7\frac{9}{10} - 6\frac{7}{10} =$

11) $8\frac{3}{4} - 5\frac{1}{12} =$

12) $6\frac{7}{8} - 3\frac{1}{8} =$

13) $7\frac{12}{35} - 3\frac{3}{7} =$

14) $6\frac{1}{3} - 4\frac{1}{9} =$

15) $10\frac{6}{7} - 7\frac{3}{7} =$

16) $9\frac{2}{3} - 3\frac{1}{3} =$

17) $5\frac{4}{11} - 3\frac{2}{11} =$

18) $7\frac{3}{10} - 5\frac{1}{5} =$

19) $8\frac{5}{6} - 5\frac{1}{12} =$

20) $3\frac{3}{4} - 3\frac{7}{16} =$

21) $8\frac{8}{13} - 3\frac{1}{3} =$

22) $6\frac{5}{6} - 4\frac{7}{30} =$

23) $5\frac{6}{7} - 4\frac{4}{11} =$

24) $6\frac{10}{19} - 2\frac{9}{19} =$

Simplify Fractions

Reduce these fractions to lowest terms

1) $\frac{18}{12} =$

2) $\frac{22}{33} =$

3) $\frac{32}{40} =$

4) $\frac{27}{36} =$

5) $\frac{8}{24} =$

6) $\frac{15}{35} =$

7) $\frac{20}{35} =$

8) $\frac{56}{70} =$

9) $\frac{9}{81} =$

10) $\frac{40}{16} =$

11) $\frac{54}{72} =$

12) $\frac{40}{120} =$

13) $\frac{12}{20} =$

14) $\frac{7}{28} =$

15) $\frac{14}{49} =$

16) $\frac{58}{87} =$

17) $\frac{72}{27} =$

18) $\frac{48}{180} =$

19) $\frac{24}{64} =$

20) $\frac{48}{42} =$

21) $\frac{120}{240} =$

22) $\frac{54}{279} =$

23) $\frac{340}{68} =$

24) $\frac{150}{600} =$

Multiplying Fractions

Find the product.

1) $\dfrac{2}{3} \times \dfrac{5}{7} =$

2) $\dfrac{3}{11} \times \dfrac{4}{9} =$

3) $\dfrac{7}{24} \times \dfrac{3}{14} =$

4) $\dfrac{7}{16} \times \dfrac{24}{35} =$

5) $\dfrac{15}{21} \times \dfrac{3}{5} =$

6) $\dfrac{18}{20} \times \dfrac{4}{9} =$

7) $\dfrac{6}{7} \times \dfrac{7}{9} =$

8) $\dfrac{54}{79} \times 0 =$

9) $\dfrac{2}{6} \times \dfrac{12}{14} =$

10) $\dfrac{24}{14} \times \dfrac{7}{8} =$

11) $\dfrac{38}{36} \times \dfrac{18}{19} =$

12) $\dfrac{8}{10} \times \dfrac{5}{64} =$

13) $\dfrac{15}{4} \times \dfrac{16}{9} =$

14) $\dfrac{25}{8} \times \dfrac{4}{10} =$

15) $\dfrac{14}{63} \times \dfrac{9}{7} =$

16) $\dfrac{12}{20} \times 4 =$

17) $\dfrac{7}{33} \times \dfrac{66}{21} =$

18) $\dfrac{5}{16} \times \dfrac{8}{10} =$

19) $\dfrac{9}{10} \times \dfrac{4}{27} =$

20) $\dfrac{6}{42} \times \dfrac{7}{12} =$

21) $\dfrac{8}{19} \times \dfrac{1}{16} =$

22) $\dfrac{10}{7} \times \dfrac{4}{80} =$

23) $\dfrac{9}{12} \times \dfrac{4}{54} =$

24) $\dfrac{60}{400} \times \dfrac{200}{600} =$

Multiplying Mixed Number

Multiply. Reduce to lowest terms.

1) $3\frac{1}{4} \times 3\frac{1}{5} =$

2) $2\frac{2}{7} \times 1\frac{1}{8} =$

3) $1\frac{1}{4} \times 2\frac{3}{5} =$

4) $2\frac{2}{9} \times 1\frac{1}{10} =$

5) $3\frac{3}{5} \times 2\frac{1}{5} =$

6) $2\frac{3}{4} \times 2\frac{2}{3} =$

7) $4\frac{1}{2} \times 1\frac{1}{9} =$

8) $2\frac{4}{5} \times 4\frac{1}{7} =$

9) $3\frac{1}{4} \times 2\frac{1}{3} =$

10) $1\frac{1}{5} \times 5\frac{1}{6} =$

11) $5\frac{1}{3} \times 2\frac{1}{8} =$

12) $2\frac{1}{3} \times 1\frac{2}{9} =$

13) $3\frac{1}{4} \times 2\frac{2}{5} =$

14) $4\frac{1}{9} \times 3\frac{1}{3} =$

15) $3\frac{1}{5} \times 2\frac{1}{7} =$

16) $4\frac{1}{2} \times 2\frac{2}{5} =$

17) $1\frac{1}{4} \times 2\frac{4}{5} =$

18) $3\frac{1}{3} \times 1\frac{1}{4} =$

19) $4\frac{4}{5} \times 1\frac{5}{7} =$

20) $6\frac{3}{4} \times 1\frac{1}{3} =$

21) $5\frac{1}{2} \times 3\frac{1}{5} =$

22) $4\frac{2}{3} \times 6\frac{1}{2} =$

Dividing Fractions

Divide these fractions.

1) $7 \div \dfrac{1}{2} =$

2) $\dfrac{9}{17} \div 9 =$

3) $\dfrac{3}{16} \div \dfrac{1}{3} =$

4) $\dfrac{9}{50} \div \dfrac{3}{10} =$

5) $\dfrac{3}{19} \div \dfrac{6}{19} =$

6) $\dfrac{2}{8} \div \dfrac{14}{32} =$

7) $0 \div \dfrac{1}{11} =$

8) $\dfrac{11}{48} \div \dfrac{22}{12} =$

9) $\dfrac{7}{24} \div \dfrac{21}{16} =$

10) $\dfrac{10}{38} \div \dfrac{5}{2} =$

11) $\dfrac{12}{25} \div \dfrac{18}{35} =$

12) $\dfrac{36}{14} \div \dfrac{18}{7} =$

13) $\dfrac{11}{27} \div \dfrac{11}{9} =$

14) $\dfrac{5}{21} \div \dfrac{30}{7} =$

15) $\dfrac{72}{37} \div \dfrac{8}{37} =$

16) $\dfrac{9}{40} \div \dfrac{54}{8} =$

17) $\dfrac{46}{9} \div \dfrac{23}{45} =$

18) $7 \div \dfrac{1}{3} =$

19) $\dfrac{63}{42} \div \dfrac{7}{6} =$

20) $\dfrac{4}{49} \div \dfrac{8}{7} =$

21) $\dfrac{9}{14} \div \dfrac{18}{21} =$

22) $\dfrac{7}{30} \div \dfrac{2}{10} =$

Common Core Math Practice Grade 5

Dividing Mixed Number

Divide the following mixed numbers. Cancel and simplify when possible.

1) $3\frac{1}{2} \div 3\frac{1}{4} =$

2) $4\frac{1}{5} \div 2\frac{1}{3} =$

3) $6\frac{3}{5} \div 8\frac{1}{4} =$

4) $5\frac{1}{2} \div 5\frac{1}{3} =$

5) $7\frac{1}{5} \div 3\frac{1}{5} =$

6) $4\frac{3}{4} \div 1\frac{9}{10} =$

7) $6\frac{1}{4} \div 2\frac{1}{2} =$

8) $3\frac{3}{5} \div 2\frac{4}{7} =$

9) $2\frac{3}{5} \div 2\frac{2}{5} =$

10) $8\frac{4}{7} \div 1\frac{1}{9} =$

11) $5\frac{1}{3} \div 6\frac{2}{5} =$

12) $2\frac{5}{9} \div 2\frac{2}{3} =$

13) $1\frac{3}{4} \div 2\frac{1}{10} =$

14) $5\frac{2}{5} \div 4\frac{1}{2} =$

15) $3\frac{4}{6} \div 2\frac{2}{7} =$

16) $3\frac{2}{3} \div 3\frac{1}{7} =$

17) $6\frac{4}{5} \div 4\frac{1}{4} =$

18) $7\frac{1}{6} \div 3\frac{1}{2} =$

19) $5\frac{5}{8} \div 3\frac{3}{4} =$

20) $7\frac{1}{7} \div 2\frac{4}{5} =$

21) $8\frac{2}{3} \div 6\frac{1}{2} =$

22) $4\frac{2}{7} \div 5\frac{5}{8} =$

23) $9\frac{3}{4} \div 2\frac{3}{5} =$

24) $7\frac{4}{5} \div 7\frac{1}{4} =$

WWW.MathNotion.com

Comparing Fractions

Compare the fractions, and write >, < or =

1) $\dfrac{13}{2}$ ____ $\dfrac{19}{10}$

2) $\dfrac{15}{4}$ ____ $\dfrac{5}{7}$

3) $\dfrac{5}{8}$ ____ $\dfrac{4}{5}$

4) $\dfrac{11}{3}$ ____ $\dfrac{12}{8}$

5) $\dfrac{2}{9}$ ____ $\dfrac{4}{7}$

6) $\dfrac{13}{5}$ ____ $\dfrac{17}{4}$

7) $\dfrac{14}{9}$ ____ $\dfrac{8}{11}$

8) $\dfrac{15}{13}$ ____ $\dfrac{21}{8}$

9) $5\dfrac{1}{10}$ ____ $8\dfrac{1}{15}$

10) $9\dfrac{1}{12}$ ____ $7\dfrac{1}{9}$

11) $4\dfrac{1}{4}$ ____ $4\dfrac{1}{7}$

12) $8\dfrac{6}{7}$ ____ $8\dfrac{3}{8}$

13) $1\dfrac{5}{9}$ ____ $4\dfrac{2}{3}$

14) $\dfrac{1}{24}$ ____ $\dfrac{2}{13}$

15) $\dfrac{42}{23}$ ____ $\dfrac{29}{72}$

16) $\dfrac{14}{200}$ ____ $\dfrac{8}{81}$

17) $19\dfrac{1}{3}$ ____ $19\dfrac{1}{7}$

18) $\dfrac{1}{8}$ ____ $\dfrac{1}{12}$

19) $\dfrac{1}{11}$ ____ $\dfrac{1}{15}$

20) $\dfrac{1}{15}$ ____ $\dfrac{7}{12}$

21) $\dfrac{10}{33}$ ____ $\dfrac{8}{59}$

22) $\dfrac{6}{7}$ ____ $\dfrac{3}{8}$

23) $6\dfrac{2}{5}$ ____ $4\dfrac{12}{5}$

24) $2\dfrac{12}{5}$ ____ $3\dfrac{4}{5}$

Answer key Chapter 4

Adding Fractions – Like Denominator

1) $\frac{2}{3}$
2) $\frac{4}{7}$
3) $\frac{7}{9}$
4) $\frac{8}{15}$
5) $\frac{9}{23}$
6) $\frac{15}{29}$
7) $\frac{8}{19}$
8) $\frac{3}{8}$
9) $\frac{13}{31}$
10) $\frac{13}{51}$
11) $\frac{4}{17}$
12) $\frac{6}{13}$
13) $\frac{24}{41}$
14) $\frac{1}{5}$
15) $\frac{13}{21}$
16) $\frac{4}{11}$
17) $\frac{6}{11}$
18) $\frac{25}{67}$
19) $\frac{9}{19}$
20) $\frac{33}{47}$

Adding Fractions – Unlike Denominator

1) $\frac{2}{3}$
2) $\frac{17}{12}$
3) $\frac{11}{10}$
4) $\frac{31}{30}$
5) $\frac{7}{15}$
6) $\frac{1}{2}$
7) $\frac{13}{21}$
8) $\frac{1}{4}$
9) $\frac{3}{4}$
10) $\frac{7}{8}$
11) $\frac{11}{21}$
12) $\frac{13}{36}$
13) $\frac{19}{35}$
14) $\frac{5}{33}$
15) $\frac{19}{27}$
16) $\frac{3}{5}$
17) $\frac{19}{45}$
18) $\frac{19}{35}$
19) $\frac{8}{11}$
20) $\frac{7}{16}$
21) $\frac{17}{48}$
22) $\frac{13}{30}$
23) $\frac{13}{42}$
24) $\frac{17}{72}$

Subtracting Fractions – Like Denominator

1) 1
2) $\frac{1}{4}$
3) $\frac{3}{13}$
4) 2
5) $\frac{2}{21}$
6) $\frac{8}{43}$
7) $\frac{6}{23}$
8) $\frac{3}{47}$
9) $\frac{1}{5}$
10) $\frac{17}{48}$
11) $\frac{4}{7}$
12) $\frac{8}{51}$
13) $\frac{4}{11}$
14) $\frac{15}{49}$
15) $\frac{6}{17}$
16) $\frac{6}{23}$
17) $\frac{4}{7}$
18) $\frac{4}{31}$

19) $\frac{9}{61}$

20) $\frac{11}{52}$

21) $\frac{5}{63}$

22) $\frac{30}{83}$

23) $\frac{24}{77}$

24) $\frac{12}{55}$

Subtracting Fractions – Unlike Denominator

1) $\frac{1}{6}$

2) $\frac{22}{45}$

3) $\frac{1}{12}$

4) $\frac{43}{90}$

5) $\frac{17}{24}$

6) $\frac{33}{80}$

7) $\frac{7}{15}$

8) $\frac{1}{30}$

9) $\frac{11}{35}$

10) $\frac{52}{99}$

11) $\frac{29}{48}$

12) $\frac{1}{20}$

13) $\frac{28}{45}$

14) $\frac{39}{70}$

15) $\frac{3}{8}$

16) $\frac{13}{63}$

17) $\frac{5}{48}$

18) $\frac{13}{40}$

19) $\frac{19}{56}$

20) $\frac{3}{11}$

Converting Mix Numbers

1) $\frac{13}{5}$

2) $\frac{19}{5}$

3) $\frac{23}{4}$

4) $\frac{29}{8}$

5) $\frac{32}{5}$

6) $\frac{71}{9}$

7) $\frac{32}{15}$

8) $\frac{40}{13}$

9) $\frac{25}{12}$

10) $\frac{35}{6}$

11) $\frac{38}{5}$

12) $\frac{39}{10}$

13) $\frac{19}{3}$

14) $\frac{35}{6}$

15) $\frac{42}{5}$

16) $\frac{35}{8}$

17) $\frac{32}{9}$

18) $\frac{31}{14}$

19) $\frac{47}{6}$

20) $\frac{41}{7}$

21) $\frac{39}{8}$

22) $\frac{29}{9}$

23) $\frac{31}{11}$

24) $\frac{75}{7}$

Converting improper Fractions

1) $3\frac{7}{16}$

2) $2\frac{27}{34}$

3) $2\frac{1}{19}$

4) $3\frac{2}{3}$

5) $3\frac{10}{17}$

6) $3\frac{29}{41}$

7) $4\frac{1}{31}$

8) $6\frac{4}{7}$

9) $3\frac{10}{11}$

10) $4\frac{2}{3}$

11) $4\frac{5}{6}$

12) $4\frac{1}{15}$

13) $2\frac{1}{3}$

14) $2\frac{1}{3}$

15) $7\frac{3}{16}$

Common Core Math Practice Grade 5

16) $9\frac{5}{6}$
17) $14\frac{2}{5}$
18) $3\frac{12}{13}$

19) $4\frac{1}{2}$
20) $9\frac{2}{3}$
21) $1\frac{2}{7}$

22) $8\frac{1}{11}$
23) $11\frac{1}{3}$
24) $9\frac{9}{19}$

Adding Mix Numbers

1) $5\frac{2}{3}$
2) $5\frac{3}{5}$
3) $3\frac{4}{7}$
4) $7\frac{1}{6}$
5) $3\frac{2}{3}$
6) $6\frac{1}{12}$
7) $6\frac{17}{18}$
8) $6\frac{1}{8}$
9) 6

10) $3\frac{7}{11}$
11) $6\frac{9}{10}$
12) $8\frac{1}{12}$
13) 9
14) $6\frac{1}{16}$
15) $6\frac{35}{36}$
16) $6\frac{23}{30}$
17) $14\frac{1}{10}$
18) $8\frac{1}{12}$

19) $9\frac{7}{9}$
20) $6\frac{13}{18}$
21) $12\frac{5}{12}$
22) $8\frac{5}{12}$
23) $9\frac{31}{56}$
24) $13\frac{1}{12}$

Subtracting Mix Numbers

1) 1
2) $\frac{2}{3}$
3) $1\frac{4}{11}$
4) $3\frac{1}{3}$
5) $3\frac{1}{8}$
6) $3\frac{19}{21}$
7) $3\frac{5}{9}$
8) $3\frac{4}{17}$

9) $6\frac{3}{7}$
10) $1\frac{1}{5}$
11) $3\frac{2}{3}$
12) $3\frac{3}{4}$
13) $3\frac{32}{35}$
14) $2\frac{2}{9}$
15) $3\frac{3}{7}$
16) $6\frac{1}{3}$

17) $2\frac{2}{11}$
18) $2\frac{1}{10}$
19) $3\frac{3}{4}$
20) $\frac{5}{16}$
21) $5\frac{11}{39}$
22) $2\frac{3}{5}$
23) $1\frac{38}{77}$
24) $4\frac{1}{19}$

Simplify Fractions

1) $\frac{3}{2}$
2) $\frac{2}{3}$
3) $\frac{4}{5}$

4) $\frac{3}{4}$
5) $\frac{1}{3}$
6) $\frac{3}{7}$
7) $\frac{4}{7}$
8) $\frac{4}{5}$
9) $\frac{1}{9}$
10) $\frac{5}{2}$

11) $\frac{3}{4}$
12) $\frac{1}{3}$
13) $\frac{3}{5}$
14) $\frac{1}{4}$
15) $\frac{2}{7}$
16) $\frac{2}{3}$
17) $\frac{8}{3}$

18) $\frac{4}{15}$
19) $\frac{3}{8}$
20) $\frac{8}{7}$
21) $\frac{1}{2}$
22) $\frac{6}{31}$
23) 5
24) $\frac{1}{4}$

Multiplying Fractions

1) $\frac{10}{21}$
2) $\frac{4}{33}$
3) $\frac{1}{16}$
4) $\frac{3}{10}$
5) $\frac{3}{7}$
6) $\frac{2}{5}$
7) $\frac{2}{3}$
8) 0

9) $\frac{2}{7}$
10) $\frac{3}{2}$
11) 1
12) $\frac{1}{16}$
13) $\frac{20}{3}$
14) $\frac{5}{4}$
15) $\frac{2}{7}$
16) $\frac{12}{5}$

17) $\frac{2}{3}$
18) $\frac{1}{4}$
19) $\frac{2}{15}$
20) $\frac{1}{12}$
21) $\frac{1}{38}$
22) $\frac{1}{14}$
23) $\frac{1}{18}$
24) $\frac{1}{20}$

Multiplying Mixed Number

1) $10\frac{2}{5}$
2) $2\frac{4}{7}$
3) $3\frac{1}{4}$
4) $2\frac{4}{9}$
5) $7\frac{23}{25}$
6) $7\frac{1}{3}$
7) 5
8) $11\frac{3}{5}$

9) $7\frac{7}{12}$
10) $6\frac{1}{5}$
11) $11\frac{1}{3}$
12) $2\frac{23}{27}$
13) $7\frac{4}{5}$
14) $13\frac{19}{27}$
15) $6\frac{6}{7}$

16) $10\frac{4}{5}$
17) $3\frac{1}{2}$
18) $4\frac{1}{6}$
19) $8\frac{8}{35}$
20) 9
21) $17\frac{3}{5}$
22) $30\frac{1}{3}$

Common Core Math Practice Grade 5

Dividing Fractions

1) 14
2) $\frac{1}{17}$
3) $\frac{9}{16}$
4) $\frac{3}{5}$
5) $\frac{1}{2}$
6) $\frac{4}{7}$
7) 0
8) $\frac{1}{8}$
9) $\frac{2}{9}$
10) $\frac{2}{19}$
11) $\frac{14}{15}$
12) 1
13) $\frac{1}{3}$
14) $\frac{1}{18}$
15) 9
16) $\frac{1}{30}$
17) 10
18) 21
19) $\frac{9}{7}$
20) $\frac{1}{14}$
21) $\frac{3}{4}$
22) $\frac{7}{6}$

Dividing Mixed Number

1) $1\frac{1}{13}$
2) $1\frac{4}{5}$
3) $\frac{4}{5}$
4) $1\frac{1}{32}$
5) $2\frac{1}{4}$
6) $2\frac{1}{2}$
7) $2\frac{1}{2}$
8) $1\frac{2}{5}$
9) $1\frac{1}{12}$
10) $7\frac{5}{7}$
11) $\frac{5}{6}$
12) $\frac{23}{24}$
13) $\frac{5}{6}$
14) $1\frac{1}{5}$
15) $1\frac{29}{48}$
16) $1\frac{1}{6}$
17) $1\frac{3}{5}$
18) $2\frac{1}{21}$
19) $1\frac{1}{2}$
20) $2\frac{27}{49}$
21) $1\frac{1}{3}$
22) $\frac{16}{21}$
23) $3\frac{3}{4}$
24) $1\frac{11}{145}$

Comparing Fractions

1) >
2) >
3) <
4) >
5) <
6) <
7) >
8) <
9) <
10) >
11) >
12) >
13) <
14) <
15) >
16) <
17) >
18) >
19) >
20) <
21) >
22) >
23) =
24) >

Chapter 5:
Decimal

Graph Decimals

Write the decimals indicated by the arrows.

1)

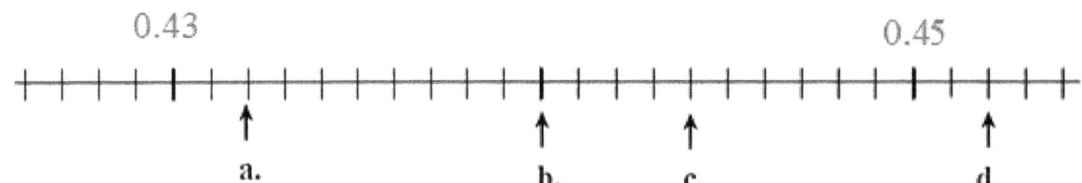

a. _____ b. _____ c. _____ d. _____

2)

a. _____ b. _____ c. _____ d. _____

3)

a. _____ b. _____ c. _____ d. _____

4)

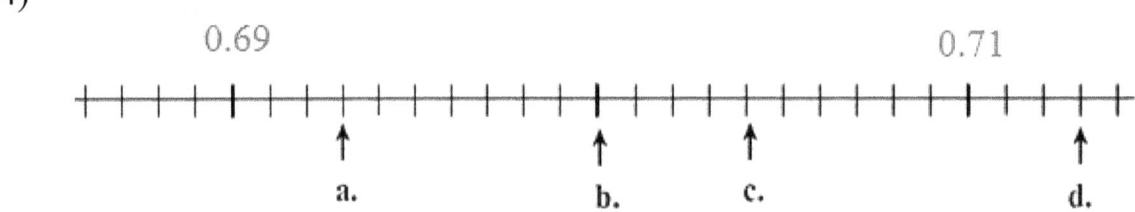

a. _____ b. _____ c. _____ d. _____

Round Decimals

Round each number to the correct place value

1) 0.9<u>2</u> =

2) 4.<u>2</u>1 =

3) 8.<u>8</u>33 =

4) 0.<u>5</u>69 =

5) <u>7</u>.779 =

6) 0.0<u>4</u>9 =

7) 9.<u>3</u>7 =

8) 25.3<u>3</u>1 =

9) 4.6<u>2</u>9 =

10) 10.<u>4</u>71 =

11) 3.<u>2</u>6 =

12) 4.<u>2</u>29 =

13) 6.<u>2</u>18 =

14) 9.1<u>6</u>24 =

15) 3<u>6</u>.29 =

16) 4<u>7</u>.68 =

17) 6<u>3</u>2.785 =

18) 577.<u>8</u>29 =

19) 24.5<u>7</u>9 =

20) 8<u>3</u>.91 =

21) 5.4<u>1</u>35 =

22) 86.<u>2</u>76 =

23) 354.<u>3</u>39 =

24) 0.9<u>2</u>66 =

25) 0.00<u>7</u>4 =

26) 8.0<u>4</u>49 =

27) 41.6<u>5</u>96 =

28) 19.0<u>8</u>27 =

Decimals Addition

Add the following.

1) 32.19 + 31.16

2) 0.54 + 0.31

3) 21.42 + 12.57

4) 57.189 + 7.221

5) 22.740 + 8.37

6) 6.712 + 4.105

7) 78.46 + 18.54

8) 66.24 + 20.36

9) 39.11 + 15.09

10) 4.88 + 19.45

11) 16.254 + 34.227

12) 36.66 + 4.82

13) 42.45 + 9.37

14) 124.41 + 5.71

Decimals Subtraction

Subtract the following

1) 8.19 − 2.46

2) 46.25 − 16.18

3) 0.89 − 0.5

4) 24.354 − 7.8

5) 0.789 − 0.06

6) 63.25 − 42.52

7) 157.75 − 94.87

8) 45.65 − 19.57

9) 65.42 − 47.71

10) 8.652 − 0.553

11) 48.61 − 29.52

12) 16.399 − 5.389

13) 35.251 − 9.169

14) 148.42 − 11.78

Decimals Multiplication

Solve.

1) 2.4 × 1.9

2) 3.6 × 5.2

3) 4.02 × 2.04

4) 45.9 × 10

5) 49.8 × 100

6) 41.56 × 4.2

7) 24.51 × 10.2

8) 1.89 × 7.35

9) 14.05 × 0.09

10) 51.03 × 4.04

11) 14.56 × 12.4

12) 9.56 × 0.04

13) 7.5 × 0.11

14) 23.1 × 4.02

Decimal Division

Dividing Decimals.

1) $7 \div 1{,}000 =$

2) $3 \div 10{,}000 =$

3) $2.8 \div 10 =$

4) $0.08 \div 100 =$

5) $5 \div 25 =$

6) $4 \div 48 =$

7) $8 \div 40 =$

8) $7 \div 140 =$

9) $9 \div 10{,}000 =$

10) $0.6 \div 0.54 =$

11) $0.4 \div 0.004 =$

12) $0.3 \div 0.15 =$

13) $0.7 \div 0.56 =$

14) $0.6 \div 0.0006 =$

15) $4.9 \div 100 =$

16) $7.3 \div 100 =$

17) $8.5 \div 10 =$

18) $16.2 \div 4.4 =$

19) $36.9 \div 3.3 =$

20) $0.8 \div 0.08 =$

21) $8.04 \div 4.2 =$

22) $0.09 \div 0.30 =$

23) $0.8 \div 6.4 =$

24) $0.07 \div 42 =$

25) $6.28 \div 0.6 =$

26) $0.026 \div 13 =$

Comparing Decimals

Write the Correct Comparison Symbol (>, < or =)

1) 1.98 ____ 3.12

2) 0.6 ____ 0.549

3) 19.01 ____ 19.010

4) 5.05 ____ 5.50

5) 0.811 ____ 0.81

6) 0.658 ____ 0.865

7) 5.46 ____ 5.391

8) 7.021 ____ 7.035

9) 56.321 ____ 56.123

10) 4.69 ____ 4.069

11) 3.55 ____ 3.555

12) 0.08 ____ 0.12

13) 2.405 ____ 2.45

14) 7.53 ____ 7.35

15) 0.11 ____ 0.011

16) 86.09 ____ 86.090

17) 0.190 ____ 0.21

18) 53.98 ____ 54.07

19) 0.072 ____ 0.720

20) 43.2 ____ 34.9

21) 17.99 ____ 20.19

22) 0.087 ____ 0.0807

23) 3.059 ____ 0.3059

24) 8.2 ____ 0.825

25) 8.77 ____ 0.877

26) 9.56 ____ 9.5600

27) 4.97 ____ 0.497

28) 3.0504 ____ 3.0540

Convert Fraction to Decimal

Write each as a decimal.

1) $\frac{6}{10} =$

2) $\frac{58}{100} =$

3) $\frac{80}{100} =$

4) $\frac{5}{40} =$

5) $\frac{4}{50} =$

6) $\frac{7}{100} =$

7) $\frac{5}{80} =$

8) $\frac{21}{84} =$

9) $\frac{36}{400} =$

10) $\frac{3}{11} =$

11) $\frac{24}{48} =$

12) $\frac{18}{48} =$

13) $\frac{6}{20} =$

14) $\frac{9}{125} =$

15) $\frac{36}{120} =$

16) $\frac{15}{20} =$

17) $\frac{73}{100} =$

18) $\frac{9}{45} =$

19) $\frac{9}{10} =$

20) $\frac{6}{60} =$

21) $\frac{7}{42} =$

22) $\frac{11}{88} =$

Answer key Chapter 5

Graph Decimals

1) a. 0.432   b. 0.44   c. 0.444   d. 0.452
2) a. 0.577   b. 0.581   c. 0.585   d. 0592
3) a. 0.714   b. 0.718   c. 0.722   d. 0.731
4) a. 0.693   b. 0.70   c. 0.704   d. 0.713

Round Decimals

1) 0.9
2) 4.2
3) 8.8
4) 0.6
5) 8.0
6) 0.05
7) 9.4
8) 25.33
9) 4.63
10) 10.5
11) 3.3
12) 4.2
13) 6.2
14) 9.16
15) 36.0
16) 48.0
17) 630.0
18) 577.8
19) 24.58
20) 84.0
21) 5.41
22) 86.3
23) 354.3
24) 0.93
25) 0.007
26) 8.04
27) 41.66
28) 19.08

Decimals Addition

1) 63.35
2) 0.85
3) 33.99
4) 64.41
5) 31.11
6) 10.817
7) 97
8) 86.6
9) 54.2
10) 24.33
11) 50.481
12) 41.48
13) 51.82
14) 130.12

Decimals Subtraction

1) 5.73
2) 30.07
3) 0.39
4) 16.554
5) 0.729
6) 20.73
7) 62.88
8) 26.08
9) 17.71
10) 8.099
11) 19.09
12) 11.01
13) 26.082
14) 136.64

Decimals Multiplication

1) 4.56
2) 18.72
3) 8.2008

Common Core Math Practice Grade 5

4) 459
5) 4,980
6) 174.552
7) 250.002
8) 13.8915
9) 1.2645
10) 206.1612
11) 180.544
12) 0.3824
13) 0.825
14) 92.862

Decimal Division

1) 0.007
2) 0.0003
3) 0.28
4) 0.0008
5) 0.2
6) 0.833….
7) 0.2
8) 0.05
9) 0.0009
10) 1.111…
11) 100
12) 2
13) 1.25
14) 1,000
15) 0.049
16) 0.073
17) 0.85
18) 3.68181…
19) 11.1818…
20) 10
21) 1.19428…
22) 0.3
23) 0.125
24) 0.00167
25) 10.467
26) 0.002

Comparing Decimals

1) <
2) >
3) =
4) <
5) >
6) <
7) >
8) <
9) >
10) >
11) <
12) <
13) <
14) >
15) >
16) =
17) <
18) <
19) <
20) >
21) <
22) >
23) >
24) >
25) >
26) =
27) >
28) <

Convert Fraction to Decimal

1) 0.6
2) 0.58
3) 0.8
4) 0.125
5) 0.08
6) 0.07
7) 0.0625
8) 0.25
9) 0.09
10) 0.27
11) 0.5
12) 0.375
13) 0.3
14) 0.072
15) 0.3

16) 0.75   19) 0.9   22) 0.125
17) 0.73   20) 0.1
18) 0.2   21) 0.167

Chapter 6: Exponent and Radicals

Common Core Math Practice Grade 5

Positive Exponents

Simplify. Your answer should contain only positive exponents.

1) $4^3 =$

2) $5^3 =$

3) $\frac{7yx}{y} =$

4) $(10x^3x)^2 =$

5) $(x^4)^3 =$

6) $(\frac{1}{2})^5 =$

7) $0^5 =$

8) $7 \times 7 \times 7 =$

9) $2 \times 2 \times 2 \times 2 \times 2 =$

10) $(3x^3y)^3 =$

11) $3^4 =$

12) $(2x^2y)^4 =$

13) $6 \times 10^3 =$

14) $0.5 \times 0.5 \times 0.5 =$

15) $\frac{1}{9} \times \frac{1}{9} \times \frac{1}{9} =$

16) $6^3 =$

17) $(4x^6y^3)^2 =$

18) $2^6 =$

19) $z \times z \times z =$

20) $9 \times 9 \times 9 \times 9 \times 9 =$

21) $(3x^3y^3z)^2 =$

22) $5^0 =$

23) $(12x^3y^{-2})^2 =$

24) $(3x^3y^2)^4 =$

WWW.MathNotion.com

Common Core Math Practice Grade 5

Negative Exponents

Simplify. Leave no negative exponents.

1) $3^{-3} =$

2) $8^{-2} =$

3) $(\frac{1}{4})^{-3} =$

4) $7^{-2} =$

5) $1^{120} =$

6) $6^{-4} =$

7) $(\frac{1}{3})^{-4} =$

8) $-12y^{-9} =$

9) $(\frac{1}{y^{-4}})^{-6} =$

10) $x^{-\frac{6}{7}} =$

11) $\frac{1}{9^{-4}} =$

12) $2^{-6} =$

13) $3^{-5} =$

14) $12^{-2} =$

15) $20^{-2} =$

16) $x^{-9} =$

17) $(x^3)^{-5} =$

18) $x^{-3} \times x^{-3} \times x^{-3} =$

19) $\frac{1}{8} \times \frac{1}{8} \times \frac{1}{8} =$

20) $10^{-4} =$

21) $200z^{-5} =$

22) $2^{-8} =$

23) $(-\frac{1}{13})^2 =$

24) $23^0 =$

25) $(\frac{1}{x})^{-31} =$

26) $16^{-2} =$

www.MathNotion.com

Add and subtract Exponents

Solve each problem.

1) $3^3 + 4^2 =$

2) $x^9 + 2x^9 =$

3) $7b^4 - 6b^4 =$

4) $8 + 6^2 =$

5) $12 - 5^2 =$

6) $18 + 2^3 =$

7) $6x^2 + 9x^2 =$

8) $7^2 + 3^4 =$

9) $4^4 - 3^3 =$

10) $9^2 - 12^0 =$

11) $8^2 - 5^2 =$

12) $11^2 + 2^4 =$

13) $13^2 - 9^2 =$

14) $10^2 + 11^2 =$

15) $5^3 - 2^4 =$

16) $1^{32} + 1^{12} =$

17) $8^2 - 7^2 =$

18) $3^5 - 3^4 =$

19) $10^2 - 2^3 =$

20) $6^2 + 3^3 =$

21) $5^3 + 2^4 =$

22) $24 + 3^2 =$

23) $15x^7 + 4x^7 =$

24) $7^0 + 9^2 =$

25) $6^2 + 6^2 =$

26) $4^2 + 2^5 =$

27) $(\frac{1}{5})^2 + (\frac{1}{5})^2 =$

28) $7^2 + 6^2 =$

Exponent multiplication

Simplify each of the following

1) $4^4 \times 4^5 =$

2) $9^2 \times 10^0 =$

3) $12^2 \times 2^2 =$

4) $a^{-7} \times a^{-7} =$

5) $y^{-6} \times y^{-6} \times y^{-6} =$

6) $4^6 \times 5^9 \times 4^{-5} \times 5^{-8} =$

7) $5x^2y^4 \times 3x^2y^5 =$

8) $(x^4)^2 =$

9) $(x^3y^4)^4 \times (x^3y^2)^{-4} =$

10) $9^5 \times 9^3 =$

11) $z^{6b} \times z^0 =$

12) $8^5 \times 8^4 =$

13) $a^{7m} \times a^{5n} =$

14) $3a^n \times 5b^n =$

15) $7^{-4} \times 6^{-4} =$

16) $9^8 \times 5^8 =$

17) $(5^7)^4 =$

18) $(\frac{1}{8})^3 \times (\frac{1}{8})^6 \times (\frac{1}{8})^4 =$

19) $(\frac{1}{6})^{41} \times 6^{41} =$

20) $(5m)^{\frac{2}{7}} \times (-4m)^{\frac{2}{7}} =$

21) $(x^3y)^{\frac{1}{3}} \times (xy^4)^{\frac{1}{3}} =$

22) $(3a^{2m}b^n)^r =$

23) $(6x^4y^4)^2 =$

24) $(x^{\frac{1}{5}}y^3)^{\frac{-1}{5}} \times (x^7y^8)^0 =$

25) $9^4 \times 9^7 =$

26) $16^{\frac{3}{8}} \times 16^{\frac{1}{8}} =$

27) $8^6 \times 2^6 =$

28) $(x^{18})^0 =$

Exponent division

Simplify. Your answer should contain only positive exponents.

1) $\dfrac{7^5}{7} =$

2) $\dfrac{21x^3}{x} =$

3) $\dfrac{a^{3m}}{a^{4n}} =$

4) $\dfrac{2x^{-7}}{14x^{-2}} =$

5) $\dfrac{54x^7}{9x^3} =$

6) $\dfrac{11x^8}{3x^9} =$

7) $\dfrac{45x^9}{15y^5} =$

8) $\dfrac{25xy^9}{x^5y^3} =$

9) $\dfrac{2x^7}{5x} =$

10) $\dfrac{36x^6y^7}{4x^7} =$

11) $\dfrac{9x^4}{27x^9y^{10}} =$

12) $\dfrac{5yx^5}{35yx^{12}} =$

13) $\dfrac{32x^4y^4}{8x^3y^5} =$

14) $\dfrac{x^{2.25}}{x^{0.25}} =$

15) $\dfrac{7x^5y}{28xy^4} =$

16) $\dfrac{36b^4r^5}{18a^3b^6} =$

17) $\dfrac{40x^5}{20x^7} =$

18) $\dfrac{24x^2}{6x^5} =$

19) $\dfrac{7^9}{7^7} =$

20) $\dfrac{2x}{x^{13}} =$

21) $\dfrac{15^9}{15^6} =$

22) $\dfrac{4xy^6}{16y^4} =$

23) $\dfrac{12x^4y}{144xy^5} =$

24) $\dfrac{42x^7}{7y^8} =$

Scientific Notation

Write each number in scientific notation.

1) 8,700,000 =

2) 6,000 =

3) 0.0005 =

4) 250,000 =

5) 0.0159 =

6) 0.056 =

7) 0.000005 =

8) 12,000,000 =

9) 800,000 =

10) 12,100,000 =

11) 0.0009 =

12) 0.000008 =

13) 0.00055 =

14) 600,000 =

15) 3,054 =

16) 1,570,000 =

17) 0.00013 =

18) 0.4 =

19) 0.97 =

20) 400,000 =

21) 4,150,000 =

22) 0.0048 =

23) 0.00027 =

24) 3,125 =

25) 63,887 =

26) 121,100 =

27) 54,200 =

28) 6,250,000 =

Square Roots

Find the square root of each number.

1) $\sqrt{49} =$

2) $\sqrt{1} =$

3) $\sqrt{25} =$

4) $\sqrt{121} =$

5) $\sqrt{400} =$

6) $\sqrt{36} =$

7) $\sqrt{0} =$

8) $\sqrt{100} =$

9) $\sqrt{16} =$

10) $\sqrt{81} =$

11) $\sqrt{144} =$

12) $\sqrt{900} =$

13) $\sqrt{64} =$

14) $\sqrt{225} =$

15) $\sqrt{9} =$

16) $\sqrt{169} =$

17) $\sqrt{1,600} =$

18) $\sqrt{0.09} =$

19) $\sqrt{625} =$

20) $\sqrt{256} =$

21) $\sqrt{1.44} =$

22) $\sqrt{4,900} =$

23) $\sqrt{484} =$

24) $\sqrt{6,400} =$

25) $\sqrt{289} =$

26) $\sqrt{2.56} =$

27) $\sqrt{361} =$

28) $\sqrt{0.01} =$

WWW.MathNotion.com

Simplify Square Roots

Simplify the following.

1) $\sqrt{80} =$

2) $\sqrt{72} =$

3) $\sqrt{20} =$

4) $\sqrt{90} =$

5) $\sqrt{288} =$

6) $\sqrt{125} =$

7) $4\sqrt{20} =$

8) $5\sqrt{200} =$

9) $\sqrt{32} =$

10) $4\sqrt{8} =$

11) $5\sqrt{7} + 6\sqrt{7} =$

12) $\frac{16}{5+\sqrt{3}} =$

13) $\sqrt{98} =$

14) $\frac{7}{4-\sqrt{3}} =$

15) $\sqrt{12} \times \sqrt{3} =$

16) $\frac{\sqrt{500}}{\sqrt{5}} =$

17) $\frac{\sqrt{60}}{\sqrt{15 \times 4}} =$

18) $\sqrt{180y^8} =$

19) $5\sqrt{49a} =$

20) $\sqrt{58+6} + \sqrt{16} =$

21) $\sqrt{162} =$

22) $\sqrt{242} =$

23) $\sqrt{175} =$

24) $\sqrt{147} =$

25) $\sqrt{512} =$

26) $\sqrt{112} =$

Answer key Chapter 6

Positive Exponents

1) 64
2) 125
3) $7x$
4) $100x^8$
5) x^{12}
6) $\frac{1}{32}$
7) 0
8) 7^3
9) 2^5
10) $27x^9y^3$
11) 81
12) $16x^8y^4$
13) 6,000
14) 0.5^3
15) $(\frac{1}{9})^3$
16) 216
17) $16x^{12}y^6$
18) 64
19) z^3
20) 9^5
21) $9x^6y^6z^2$
22) 1
23) $\frac{144x^6}{y^4}$
24) $81x^{12}y^8$

Negative Exponents

1) $\frac{1}{27}$
2) $\frac{1}{64}$
3) 64
4) $\frac{1}{49}$
5) 1
6) $\frac{1}{1,296}$
7) 81
8) $\frac{-12}{y^9}$
9) $\frac{1}{y^{24}}$
10) $\frac{1}{x^7}$
11) 9^4
12) $\frac{1}{64}$
13) $\frac{1}{243}$
14) $\frac{1}{144}$
15) $\frac{1}{400}$
16) $\frac{1}{x^9}$
17) $\frac{1}{x^{15}}$
18) $\frac{1}{x^9}$
19) $\frac{1}{8^3}$
20) $\frac{1}{10,000}$
21) $\frac{200}{z^5}$
22) $\frac{1}{256}$
23) $\frac{1}{169}$
24) 1
25) x^{31}
26) $\frac{1}{256}$

Add and subtract Exponents

1) 43
2) $3x^9$
3) b^4
4) 44
5) -13
6) 26
7) $15x^2$
8) 130
9) 229
10) 80
11) 39
12) 137
13) 88
14) 221
15) 109

16) 2
17) 15
18) 162
19) 92
20) 63
21) 141
22) 33
23) $19x^7$
24) 82
25) 72
26) 48
27) $\frac{2}{25}$
28) 85

Exponent multiplication

1) 4^9
2) 81
3) 576
4) a^{-14}
5) y^{-18}
6) 20
7) $15x^4y^9$
8) x^8
9) y^8
10) 9^8
11) Z^{6b}
12) 8^9
13) a^{7m+5n}
14) $15(ab)^n$
15) 42^{-4}
16) 45^8
17) 5^{28}
18) $(\frac{1}{8})^{13}$
19) 1
20) $(-20m^2)^{\frac{2}{7}}$
21) $x^{\frac{10}{3}}y^{\frac{5}{3}}$
22) $3^r a^{2mr} b^{nr}$
23) $36x^8 y^8$
24) $x^{\frac{-1}{25}} y^{\frac{-3}{5}}$
25) 9^{11}
26) $16^{\frac{1}{2}}$
27) $16^6 = 2^{24}$
28) 1

Exponent division

1) 7^4
2) $21x^2$
3) a^{3m-4n}
4) $\frac{1}{7x^5}$
5) $6x^4$
6) $\frac{11}{3x}$
7) $\frac{3x^9}{y^5}$
8) $\frac{25y^6}{x^4}$
9) $\frac{2x^6}{5}$
10) $\frac{9y^7}{x}$
11) $\frac{1}{3x^5 y^{10}}$
12) $\frac{1}{7x^7}$
13) $\frac{4x}{y}$
14) x^2
15) $\frac{x^4}{4y^3}$
16) $\frac{2r^5}{a^3 b^2}$
17) $\frac{2}{x^2}$
18) $\frac{4}{x^3}$
19) 7^2
20) $\frac{2}{x^{12}}$
21) 15^3
22) $\frac{1}{4}xy^2$
23) $\frac{x^3}{12y^4}$
24) $\frac{6x^7}{y^8}$

Scientific Notation

1) 8.7×10^6
2) 6×10^3
3) 5×10^{-4}
4) 2.5×10^5
5) 1.59×10^{-2}
6) 5.6×10^{-2}

7) 5×10^{-6}
8) 1.2×10^7
9) 8×10^5
10) 1.21×10^7
11) 9×10^{-4}
12) 8×10^{-6}
13) 5.5×10^{-4}
14) 6×10^5

15) 3.054×10^3
16) 1.57×10^6
17) 1.3×10^{-4}
18) 4×10^{-1}
19) 9.7×10^{-2}
20) 4×10^5
21) 4.15×10^6
22) 4.8×10^{-3}

23) 2.7×10^{-4}
24) 3.125×10^3
25) 6.3887×10^4
26) 1.211×10^5
27) 5.42×10^4
28) 6.25×10^6

Square Roots

1) 7
2) 1
3) 5
4) 11
5) 20
6) 6
7) 0
8) 10
9) 4
10) 9

11) 12
12) 30
13) 8
14) 15
15) 3
16) 13
17) 40
18) 0.3
19) 25
20) 16

21) 1.2
22) 70
23) 22
24) 80
25) 17
26) 1.6
27) 19
28) 0.1

Simplify Square Roots

1) $4\sqrt{5}$
2) $6\sqrt{2}$
3) $2\sqrt{5}$
4) $3\sqrt{10}$
5) $12\sqrt{2}$
6) $5\sqrt{5}$
7) $8\sqrt{5}$
8) $50\sqrt{2}$
9) $4\sqrt{2}$

10) $8\sqrt{2}$
11) $11\sqrt{7}$
12) $\frac{8}{11}(5 - \sqrt{3})$
13) $7\sqrt{2}$
14) $4 - \sqrt{3}$
15) 6
16) 10
17) 1
18) $6y^4\sqrt{5}$

19) $35\sqrt{a}$
20) 12
21) $9\sqrt{2}$
22) $11\sqrt{2}$
23) $5\sqrt{7}$
24) $7\sqrt{3}$
25) $16\sqrt{2}$
26) $4\sqrt{7}$

Chapter 7: Ratio, Proportion and Percent

Proportions

Find a missing number in a proportion.

1) $\dfrac{5}{3} = \dfrac{35}{a}$

2) $\dfrac{a}{6} = \dfrac{16}{24}$

3) $\dfrac{11}{44} = \dfrac{a}{4}$

4) $\dfrac{14}{a} = \dfrac{70}{30}$

5) $\dfrac{5}{a} = \dfrac{15}{54}$

6) $\dfrac{\sqrt{4}}{5} = \dfrac{a}{60}$

7) $\dfrac{3}{9} = \dfrac{12}{a}$

8) $\dfrac{8}{16} = \dfrac{a}{31}$

9) $\dfrac{8}{a} = \dfrac{5.6}{7}$

10) $\dfrac{3}{16} = \dfrac{9}{a}$

11) $\dfrac{18}{9} = \dfrac{6}{a}$

12) $\dfrac{15}{a} = \dfrac{5}{25}$

13) $\dfrac{6}{7} = \dfrac{a}{13}$

14) $\dfrac{\sqrt{49}}{4} = \dfrac{42}{a}$

15) $\dfrac{7}{a} = \dfrac{7.7}{46.2}$

16) $\dfrac{80}{170} = \dfrac{a}{340}$

17) $\dfrac{32}{100} = \dfrac{a}{55}$

18) $\dfrac{43}{129} = \dfrac{a}{3}$

19) $\dfrac{20}{18} = \dfrac{3}{a}$

20) $\dfrac{8}{9} = \dfrac{64}{a}$

Reduce Ratio

Reduce each ratio to the simplest form.

1) 6: 24 =

2) 7: 42 =

3) 72: 40 =

4) 30: 25 =

5) 11: 110 =

6) 24: 3 =

7) 60: 300 =

8) 3: 108 =

9) 15: 45 =

10) 5.4: 6.3 =

11) 220: 660 =

12) 6: 10 =

13) 150: 250 =

14) 32: 48 =

15) 42: 84 =

16) 36: 9 =

17) 180: 45 =

18) 30: 300 =

19) 84: 56 =

20) 104: 132 =

21) 45: 90 =

22) 48: 56 =

23) 4: 60 =

24) 16: 96 =

Word Problems

Solve each word problem.

1) Bob has 18 red cards and 45 green cards. What is the ratio of Bob's red cards to his green cards? _____

2) In a party, 15 soft drinks are required for every 18 guests. If there are 378 guests, how many soft drinks is required? _____

3) In Bob's class, 45 of the students are tall and 25 are short. In Mason's class 117 students are tall and 65 students are short. Which class has a higher ratio of tall to short students? _____

4) The price of 3 apples at the Quick Market is $1.44. The price of 11 of the same apples at Walmart is $5.50. Which place is the better buy? _____

5) The bakers at a Bakery can make 300 bagels in 6 hours. How many bagels can they bake in 15 hours? What is that rate per hour? _____

6) You can buy 18 cans of green beans at a supermarket for $11.30. How much does it cost to buy 54 cans of green beans? _____

7) The ratio of boys to girls in a class is 7:5. If there are 21 boys in the class, how many girls are in that class? _____

8) The ratio of red marbles to blue marbles in a bag is 3:7. If there are 60 marbles in the bag, how many of the marbles are red? _____

Percent

Find the Percent of Numbers.

1) 20% of 40 =

2) 24% of 18 =

3) 12% of 15 =

4) 14% of 50 =

5) 7% of 70 =

6) 25% of 22 =

7) 15% of 45 =

8) 10% of 56 =

9) 30% of 74 =

10) 2.5% of 60 =

11) 75% of 16 =

12) 40% of 70 =

13) 14% of 120 =

14) 3% of 200 =

15) 35% of 0 =

16) 20% of 110 =

17) 32% of 55 =

18) 30% of 80 =

19) 9% of 14 =

20) 7% of 12 =

21) 18% of 46 =

22) 70% of 15 =

23) 12% of 55 =

24) 6% of 130 =

25) 80% of 260 =

26) 3% of 7 =

27) 11% of 420 =

28) 20% of 56 =

Convert Fraction to Percent

Write each as a percent.

1) $\frac{1}{5} =$

2) $\frac{2}{5} =$

3) $\frac{8}{16} =$

4) $\frac{3}{7} =$

5) $\frac{6}{14} =$

6) $\frac{13}{52} =$

7) $\frac{16}{22} =$

8) $\frac{21}{45} =$

9) $\frac{9}{75} =$

10) $\frac{3}{12} =$

11) $\frac{8}{24} =$

12) $\frac{36}{10} =$

13) $\frac{12}{40} =$

14) $\frac{32}{50} =$

15) $\frac{8}{29} =$

16) $\frac{3}{33} =$

17) $\frac{16}{44} =$

18) $\frac{14}{24} =$

19) $\frac{16}{86} =$

20) $\frac{9}{90} =$

21) $\frac{12}{300} =$

22) $\frac{560}{280} =$

WWW.MathNotion.com

Convert Decimal to Percent

Write each as a percent.

1) 0.159 =

2) 0.24 =

3) 3.1 =

4) 0.042 =

5) 0.007 =

6) 0.549 =

7) 0.157 =

8) 0.69 =

9) 0.004 =

10) 0.311 =

11) 0.782 =

12) 25.3 =

13) 7.585 =

14) 0.3 =

15) 4.82 =

16) 0.0254 =

17) 0.0079 =

18) 0.51 =

19) 4.15 =

20) 0.531 =

21) 6.323 =

22) 0.154 =

23) 3.9 =

24) 0.6 =

25) 1.7 =

26) 57.4 =

27) 3.05 =

28) 0.002 =

Answer key Chapter 7

Proportions

1) $a = 21$
2) $a = 4$
3) $a = 1$
4) $a = 6$
5) $a = 18$
6) $a = 24$
7) $a = 36$
8) $a = 15.5$
9) $a = 10$
10) $a = 48$
11) $a = 3$
12) $a = 75$
13) $a = \frac{78}{7}$
14) $a = 24$
15) $a = 42$
16) $a = 160$
17) $a = 17.6$
18) $a = 1$
19) $a = 2.7$
20) $a = 72$

Reduce Ratio

1) 1: 4
2) 1: 6
3) 9: 5
4) 6: 5
5) 1: 10
6) 8: 1
7) 1: 5
8) 1: 36
9) 1: 3
10) 0.6: 0.7
11) 11: 33
12) 0.6: 1
13) 3: 5
14) 2: 3
15) 1: 2
16) 4: 1
17) 4: 1
18) 1: 10
19) 3: 2
20) 26: 33
21) 1: 2
22) 6: 7
23) 1: 15
24) 1: 6

Word Problems

1) 2: 5
2) 315
3) The ratio for both classes is 9 to 5.
4) Quick Market is a better buy.
5) 750, the rate is 50per hour.
6) $33.90
7) 15
8) 18

Percent

1) 8
2) 4.32
3) 1.8
4) 7
5) 4.9
6) 5.5
7) 6.75
8) 5.6
9) 22.2
10) 1.5
11) 12
12) 28

Common Core Math Practice Grade 5

13) 16.8
14) 6
15) 0
16) 22
17) 17.6
18) 24

19) 1.26
20) 0.84
21) 8.28
22) 10.5
23) 6.6
24) 7.8

25) 208
26) 0.21
27) 46.2
28) 11.2

Convert Fraction to Percent

1) 20%
2) 40%
3) 50%
4) 42.86%
5) 29.31%
6) 25%
7) 72.73%
8) 46.67%

9) 12%
10) 25%
11) 33.33%
12) 360%
13) 30%
14) 64%
15) 27.59%
16) 9.09%

17) 36.36%
18) 58.33%
19) 18.6%
20) 10%
21) 4%
22) 200%

Convert Decimal to Percent

1) 15.9%
2) 24%
3) 310%
4) 4.2%
5) 0.7%
6) 54.9%
7) 15.7%
8) 69%
9) 0.4%
10) 31.1%

11) 78.2%
12) 2,530%
13) 758.5%
14) 30%
15) 482%
16) 2.54%
17) 0.79%
18) 51%
19) 415%
20) 53.1%

21) 632.3%
22) 15.4%
23) 390%
24) 60%
25) 170%
26) 5,740%
27) 305%
28) 0.2%

Chapter 8:

Measurement

Reference Measurement

LENGTH	
Customary	Metric
1 mile (mi) = 1,760 yards (yd)	1 kilometer (km) = 1,000 meters (m)
1 yard (yd) = 3 feet (ft)	1 meter (m) = 100 centimeters (cm)
1 foot (ft) = 12 inches (in.)	1 centimeter(cm) = 10 millimeters(mm)
VOLUME AND CAPACITY	
Customary	Metric
1 gallon (gal) = 4 quarts (qt)	1 liter (L) = 1,000 milliliters (mL)
1 quart (qt) = 2 pints (pt.)	
1 pint (pt.) = 2 cups (c)	
1 cup (c) = 8 fluid ounces (Fl oz)	
WEIGHT AND MASS	
Customary	Metric
1 ton (T) = 2,000 pounds (lb.)	1 kilogram (kg) = 1,000 grams (g)
1 pound (lb.) = 16 ounces (oz)	1 gram (g) = 1,000 milligrams (mg)
Time	
1 year = 12 months	
1 year = 52 weeks	
1 week = 7 days	
1 day = 24 hours	
1 hour = 60 minutes	
1 minute = 60 seconds	

Metric Length Measurement

Convert to the units.

1) 2,000 mm = _____ cm

2) 5 m = _____ mm

3) 7 m = _____ cm

4) 9 km = _____ m

5) 5,000 mm = _____ m

6) 2,800 cm = _____ m

7) 13 m = _____ cm

8) 4,000 mm = _____ cm

9) 20,000 mm = _____ m

10) 7 km = _____ mm

11) 6 km = _____ m

12) 3 m = _____ cm

13) 17,000 m = _____ km

14) 500,000 m = _____ km

Customary Length Measurement

Convert to the units.

1) 15 ft = _____ in

2) 8 ft = _____ in

3) 7 yd = _____ ft

4) 9 yd = _____ ft

5) 3 yd = _____ in

6) 3 mi = _____ in

7) 7,200 in = _____ yd

8) 252 in = _____ yd

9) 8,800 yd = _____ mi

10) 12 yd = _____ in

11) 4 mi = _____ yd

12) 47,520 ft = _____ mi

13) 60 in = _____ ft

14) 25 yd = _____ ft

15) 36 in = _____ ft

16) 2 mi = _____ ft

Common Core Math Practice Grade 5

Metric Capacity Measurement

Convert the following measurements.

1) 50 l = _____ ml

2) 4 l = _____ ml

3) 13 l = _____ ml

4) 8 l = _____ ml

5) 19 l = _____ ml

6) 2 l = _____ ml

7) 70,000 ml = _____ l

8) 8,000 ml = _____ l

9) 37,000 ml = _____ l

10) 200,000 ml = _____ l

11) 6,000,000 ml = _____ l

12) 40,000 ml = _____ l

Customary Capacity Measurement

Convert the following measurements.

1) 2 gal = _____ qt.

2) 11 gal = _____ pt.

3) 3 gal = _____ c.

4) 14 pt. = _____ c

5) 43 c = _____ fl oz

6) 16 qt = _____ pt.

7) 8 qt = _____ c

8) 29 pt. = _____ c

9) 6,720 c = _____ gal

10) 144 pt. = _____ gal

11) 72 qt = _____ gal

12) 92 pt. = _____ qt

13) 4,600 c = _____ qt

14) 146 c = _____ pt.

15) 108 qt = _____ gal

16) 1,848 pt. = _____ qt

17) 31 gal = _____ pt.

18) 6 qt = _____ c

19) 640 c = _____ gal

20) 104 fl oz = _____ c

www.MathNotion.com

Metric Weight and Mass Measurement

Convert.

1) 7 kg = _____ g

2) 3 kg = _____ g

3) 13 kg = _____ g

4) 21 kg = _____ g

5) 9 kg = _____ g

6) 121 kg = _____ g

7) 249 kg = _____ g

8) 4,000 g = _____ kg

9) 6,000 g = _____ kg

10) 17,000 g = _____ kg

11) 129,000 g = _____ kg

12) 220,000 g = _____ kg

13) 9,000,000 g = _____ kg

14) 11,000,000 g = _____ kg

Customary Weight and Mass Measurement

Convert.

1) 16,000 lb. = _____ T

2) 20,000 lb. = _____ T

3) 170,000 lb. = _____ T

4) 44,000 lb. = _____ T

5) 7 lb. = _____ oz

6) 4 lb. = _____ oz

7) 10 lb. = _____ oz

8) 24 T = _____ lb.

9) 3 T = _____ lb.

10) 9 T = _____ lb.

11) 112 T = _____ lb.

12) 2 T = _____ oz

13) 5 T = _____ oz

14) 224 oz = _____ lb.

Time

Convert to the units.

1) 16 hr. = _____ min

2) 9 year = _____ week

3) 2 hr. = _____ sec

4) 12 min = _____ sec

5) 600 min = _____ hr

6) 730 day = _____ year

7) 3 year = _____ hr.

8) 27 day = _____ hr

9) 4 day = _____ min

10) 540 min = _____ hr

11) 16 year = _____ month

12) 19,200 sec = _____ min

13) 288 hr = _____ day

14) 23 weeks = _____ day

How much time has passed?

15) From 4:15 A.M. to 7:25 A.M.: ____ hours and ____ minutes.

16) From 3:40 A.M. to 8:25 A.M.: ____ hours and ____ minutes.

17) It's 8:50 P.M. What time was 2 hours ago? _____ O'clock

18) 3:20 A.M to 6:40 AM: _____ hours and _____ minutes.

19) 3:30 A.M to 6:05 AM: _____ hours and _____ minutes.

20) 7:10 A.M. to 8:15 AM. = _____ hour(s) and _____ minutes.

21) 11:55 A.M. to 4:25 PM. = _____ hour(s) and _____ minutes

22) 7:18 A.M. to 7:52 A.M. = _____ minutes

23) 9:13 A.M. to 9:50 A.M. = _____ minutes

Answers of Worksheets – Chapter 8

Metric length

1) 200 cm
2) 5,000 mm
3) 700 cm
4) 9,000 m
5) 5 m
6) 28 m
7) 1,300 cm
8) 400 cm
9) 20 m
10) 7,000,000 mm
11) 6,000 m
12) 300 cm
13) 17 km
14) 500 km

Customary Length

1) 180
2) 96
3) 21
4) 27
5) 108
6) 190,080
7) 200
8) 7
9) 5
10) 432
11) 7,040
12) 9
13) 5
14) 75
15) 3
16) 10,560

Metric Capacity

1) 50,000 ml
2) 4,000 ml
3) 13,000 ml
4) 8,000 ml
5) 19,000 ml
6) 2,000 ml
7) 70 L
8) 8 L
9) 37 L
10) 200 L
11) 6,000 L
12) 40 L

Customary Capacity

1) 8 qt
2) 88 pt.
3) 48 c
4) 28 c
5) 344 fl oz
6) 12 pt.
7) 32 c
8) 58 c
9) 420 gal
10) 18 gal
11) 18 gal
12) 46 qt
13) 1,150 qt
14) 73 pt.
15) 27 gal
16) 927 qt
17) 248 pt.
18) 24 c
19) 40 gal
20) 13 c

Metric Weight and Mass

1) 7,000 g
2) 3,000 g
3) 13,000 g
4) 21,000 g
5) 9,000 g
6) 121,000 g
7) 249,000 g
8) 4 kg
9) 6 kg
10) 17 kg
11) 129 kg
12) 220 kg
13) 9,000 kg
14) 11,000 kg

Customary Weight and Mass

1) 8 T
2) 10 T
3) 85 T
4) 22 T
5) 112 oz
6) 64 oz
7) 160 oz
8) 48,000 lb.
9) 6,000 lb.
10) 18,000 lb.
11) 224,000 lb.
12) 64,000 oz
13) 160,000 oz
14) 14 lb

Time

1) 960 min
2) 468 weeks
3) 7,200 sec
4) 720 sec
5) 10 hr
6) 2 year
7) 26,280 hr
8) 648 hr
9) 5,760 min
10) 9 hr
11) 192 months
12) 320 min
13) 12 days
14) 161 days
15) 3:10
16) 4:45
17) 6:50 P.M.
18) 3:20
19) 2:35
20) 1:05
21) 4:30
22) 34 minutes
23) 37 minutes

Chapter 9: Algebraic Expressions

Find a Rule

Complete the output.

1- **Rule:** the output is $x + 15$

Input	x	7	11	18	26	34
Output	y					

2- **Rule:** the output is $x \times 12$

Input	x	2	5	9	13	17
Output	y					

3- **Rule:** the output is $x \div 8$

Input	x	104	184	136	312	440
Output	y					

Find a rule to write an expression.

4- **Rule:** _____

Input	x	12	14	17	21
Output	y	72	84	102	126

5- **Rule:** _____

Input	x	9	18	27	39
Output	y	17	26	33	47

6- **Rule:** _____

Input	x	112	168	224	357
Output	y	16	24	32	51

Variables and Expressions

Write a verbal expression for each algebraic expression.

1) $3a - 7b$

2) $5c^2 + 9d$

3) $x - 13$

4) $\frac{90}{17}$

5) $m^2 + n^3$

6) $6x + 3$

7) $a^2 - 11b + 6$

8) $x^3 + 8y^2 - 6$

9) $\frac{1}{6}x + \frac{3}{4}y - 9$

10) $\frac{1}{3}(x + 7) - 12y$

Write an algebraic expression for each verbal expression.

11) 5 less than y

12) The product of 6 and a

13) The 32 divided by z

14) The product of 4 and the third power of a

15) 6 more than h to the seventh power

16) 43 more than triple d

17) One fifth the square of x

18) The difference of 34 and 6 times a number

19) 57 more than the cube of a number

20) One-quarters the cube of a number

Translate Phrases

Write an algebraic expression for each phrase.

1) A number increased by fifty–four.

2) The sum of ninety and 4 times a number

3) The difference between seventy–eight and a number.

4) The quotient of sixty-one and a number.

5) Triple a number decreased by 80.

6) two times the sum of a number and – 45.

7) A number divided by – 18.

8) The quotient of 38 and the product of a number and – 12.

9) nine subtracted from 4 times a number.

10) The difference of six and a number.

Distributive Property

Multiply using the distributive property.

1) $3(x + 3) =$ _____

2) $4(x + 11) =$ _____

3) $(x + 9)5 =$ _____

4) $7(x + 6) =$ _____

5) $8(x + 8) =$ _____

6) $10(x + 4) =$ _____

7) $9(x + 10) =$ _____

8) $4(x + 8) =$ _____

9) $11(x + 6) =$ _____

10) $(x + 7)6 =$ _____

11) $(x + 13)4 =$ _____

12) $3(x + 12) =$ _____

13) $2(9x - 4) =$ _____

14) $7(8x - 3) =$ _____

15) $8(9x - 5) =$ _____

16) $(4x - 2)3 =$ _____

17) $(7x - 2)7 =$ _____

18) $(2x - 3)12 =$ _____

19) $3(4x - 1) =$ _____

20) $(-2)(4x - 3) =$ _____

21) $(-5)(x - 9) =$ _____

22) $(-7)(3x - 1) =$ _____

23) $(5x + 2)(-9) =$ _____

24) $(x + 6)(-12) =$ _____

Evaluate One Variable Expressions

Evaluate each using the values given.

1) $x + 2x, x = 5$

2) $3(-7 + x), x = 2$

3) $6x + 4x, x = -1$

4) $2(6 - x) + 8, x = 4$

5) $5x + 4x - 15, x = 3$

6) $3x + 7x + 22, x = -2$

7) $4x - 3x - 6, x = 4$

8) $\frac{2(2x+10)}{4}, x = 3$

9) $3x - 87, x = 25$

10) $\frac{x}{23}, x = 138$

11) $3(4 + 3x) - 24, x = 3$

12) $8(x + 2) - 31, x = 5$

13) $\frac{x+(-7)}{-2}, x = -7$

14) $6(5 - 2x) + 8, x = 3$

15) $-12 - \frac{x}{4} + 6x, x = 20$

16) $3x + 13x, x = 2$

17) $-10x + 2(6 + 2x), x = -5$

18) $3x + 15x, x = 3$

19) $\frac{(3x-7)}{5}, x = 14$

20) $4(-2 - 3x), x = 7$

21) $6x - (6 - x), x = 2$

22) $\left(-\frac{30}{x}\right) + 3 + x, x = 6$

23) $-\frac{x \times 8}{x}, x = 8$

24) $3(-2 - 4x), x = 3$

25) $3x^2 + 5x, x = 2$

26) $7(4x + 2) - 3(x - 8), x = 1$

27) $-4x - 3, x = -6$

28) $6x + 3x, x = 2$

Answer key Chapter 9

Find a Rule

1)
Input	x	7	11	18	26	34
Output	y	22	26	33	41	49

2)
Input	x	2	5	9	13	17
Output	y	24	60	108	156	204

3)
Input	x	104	184	136	312	440
Output	y	13	23	17	39	55

4) y = 6x   5) y = x + 8   6) y = x ÷ 7

Variables and Expressions

1) 3 times a minus 7 times b
2) 5 times c squared plus 9 times d
3) a number minus 13
4) the quotient of 90 and 17
5) m squared plus n cubed
6) the product of 6 and x plus 3
7) a squared minus the product of 11 and b plus 6
8) x cubed plus the product of 8 and y squared minus 6
9) the sum of one–sixth of x and three–quarters of y, minus 9
10) one–fifth of the sum of x and 7 minus the product of 12 and y

11) 5 <y
12) 6a
13) $\frac{32}{z}$
14) $4a^3$
15) 6 >h^7
16) 3d< 43
17) $\frac{1}{5}x^2$
18) 34 – 6a
19) 57 >a^3
20) $\frac{1}{4}x^3$

Translate Phrases

1) x + 54
2) 90 + 4x
3) 78 – x
4) $\frac{61}{x}$
5) 3x– 80
6) 2(x + (–45))
7) $\frac{x}{-18}$
8) $\frac{38}{-12x}$
9) 4x – 9
10) 6 – x

Distributive Property

1) $3x + 9$
2) $4x + 44$
3) $5x + 45$
4) $7x + 42$
5) $8x + 64$
6) $10x + 40$

7) $9x + 90$
8) $4x + 32$
9) $11x + 66$
10) $6x + 42$
11) $4x + 52$
12) $3x + 36$

13) $18x - 8$
14) $56x - 21$
15) $72x - 40$
16) $12x - 6$
17) $49x - 14$
18) $24x - 36$

19) $12x - 3$
20) $-8x + 6$
21) $-5x + 45$
22) $-21x + 7$
23) $-45x - 18$
24) $-12x - 72$

Evaluate One Variable Expressions

1) 15
2) −15
3) −10
4) 12
5) 12
6) 2
7) −2

8) 8
9) −12
10) 6
11) 15
12) 25
13) 7
14) 2

15) 103
16) 32
17) 42
18) 54
19) 7
20) −92
21) 8

22) 4
23) −8
24) −42
25) 22
26) 63
27) 21
28) 18

Chapter 10: Symmetry and Transformations

Line Segments

Write each as a line, ray, or line segment.

1)

2)

3)

4)

5)

6)

7)

8)

Parallel, Perpendicular and Intersecting Lines

State whether the given pair of lines are parallel, perpendicular, or intersecting.

1)

2)

3)

4)

5)

6)

7)

8)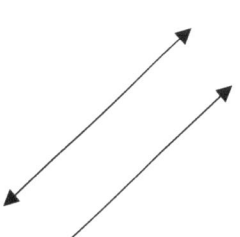

Identify Lines of Symmetry

Tell whether the line on each shape a line of symmetry is.

1)

2)

3)

4)

5)

6)

7)

8)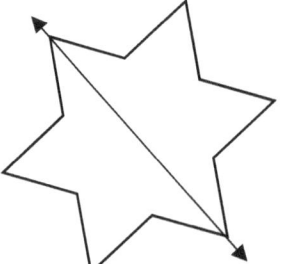

Lines of Symmetry

Draw lines of symmetry on each shape. Count and write the lines of symmetry you see.

1)

2)

3)

4)

5)

6)

7)

8)

Identify Three–Dimensional Figures

Write the name of each shape.

1)

2)

3)

4)

5)

6)

7)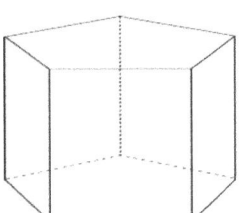

Vertices, Edges, and Faces

Complete the chart below.

	Shape	Number of edges	Number of faces	Number of vertices
1)	(hexagonal prism)	___	___	___
2)	(rectangular prism)	___	___	___
3)	(tetrahedron)	___	___	___
4)	(square pyramid)	___	___	___
5)	(cube)	___	___	___
6)	(rectangular prism)	___	___	___

Identify Faces of Three-Dimensional Figures

Write the number of faces.

1)

2)

3)

4)

5)

6)

7)

8)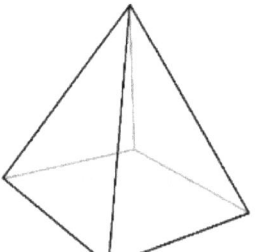

Answers of Worksheets – Chapter 10

Line Segments

1) Line
2) Line segment
3) Line segment
4) Ray
5) Line
6) Ray
7) Line segment
8) Ray

Parallel, Perpendicular and Intersecting Lines

1) Perpendicular
2) Parallel
3) Intersection
4) Perpendicular
5) Intersection
6) Parallel
7) Perpendicular
8) Parallel

Identify lines of symmetry

1) No
2) yes
3) yes
4) No
5) yes
6) No
7) No
8) yes

lines of symmetry

1)

2)

3)

4)

5)   6)

7)   8)

Identify Three–Dimensional Figures

1) Square pyramid
2) Triangular prism
3) Triangular pyramid
4) Cube
5) Hexagonal prism
6) Rectangular prism
7) Pentagonal prism

Vertices, Edges, and Faces

	Shape	Number of edges	Number of faces	Number of vertices
1)		18	8	12
2)		12	6	8
3)		6	4	4

4)   8   5   5

5)   12   6   8

6)   15   7   10

Identify Faces of Three–Dimensional Figures

1) 4   4) 2   7) 6
2) 6   5) 7   8) 5
3) 8   6) 5

Chapter 11:

Geometry

Area and Perimeter of Square

Find the perimeter and area of each squares.

1)

Perimeter: .... :

Area: .... :

2)

Perimeter: .... :

Area: .... :

3)

Perimeter: .... :

Area: .... :

4)

Perimeter: .... :

Area: .... :

5)

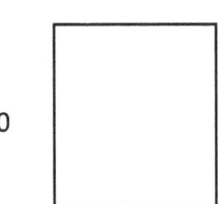

Perimeter: .... :

Area: .... :

6)

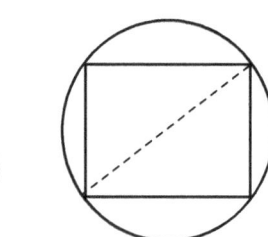

Perimeter of Square: .... :

Area of Square: .... :

Area and Perimeter of Rectangle

Find the perimeter and area of each rectangle.

1)

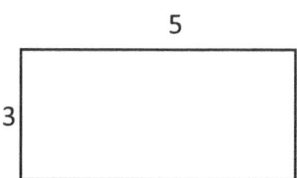

Perimeter: _____:

Area: _____:

2)

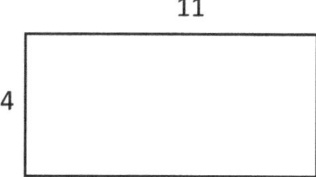

Perimeter: _____:

Area: _____:

3)

```
          10
     ┌─────────┐
 1.2 │         │
     └─────────┘
```

Perimeter: _____:

Area: _____:

4)

```
          9 1/2
     ┌─────────┐
   2 │         │
     └─────────┘
```

Perimeter: _____:

Area: _____:

5)

```
          7.2
     ┌─────────┐
   5 │         │
     └─────────┘
```

Perimeter: _____:

Area: _____:

6)

```
          8.4
     ┌─────────┐
 3.6 │         │
     └─────────┘
```

Perimeter: _____:

Area: _____:

Area and Perimeter of Triangle

Find the perimeter and area of each triangle.

1)

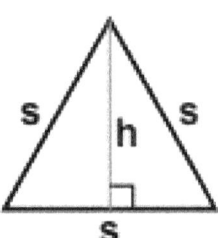

Perimeter: _____ :

Area: _____ :

2)

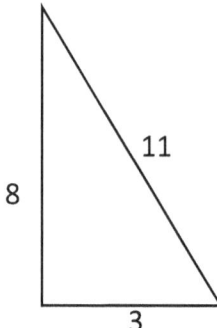

Perimeter: _____ :

Area: _____ :

3)

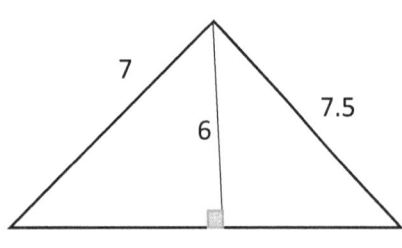

Perimeter: _____ :

Area: _____ :

4)

$s=10$

$h=6.4$

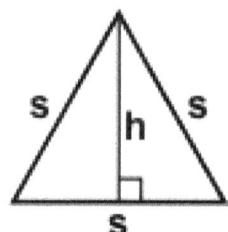

Perimeter: _____ :

Area: _____ :

5)

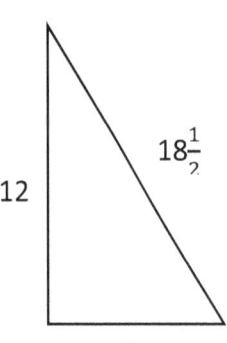

Perimeter: _____ :

Area: _____ :

6)

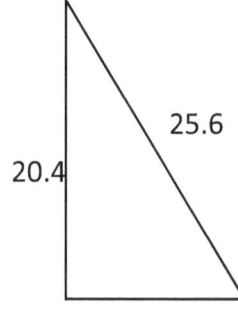

Perimeter: _____ :

Area: _____ :

Area and Perimeter of Trapezoid

Find the perimeter and area of each trapezoid.

1)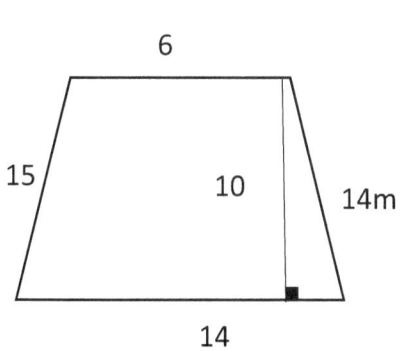

Perimeter: ....:

Area: ....:

2)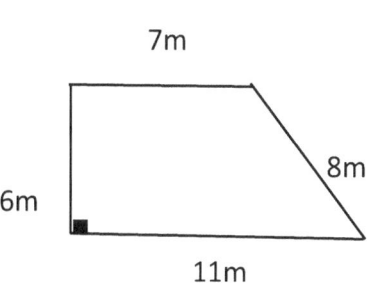

Perimeter: ....:

Area: ....:

3)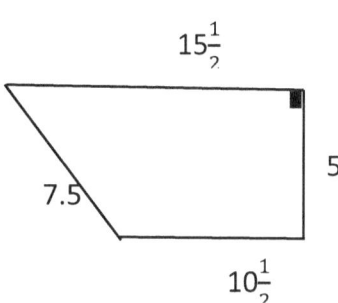

Perimeter: ....:

Area: ....:

4)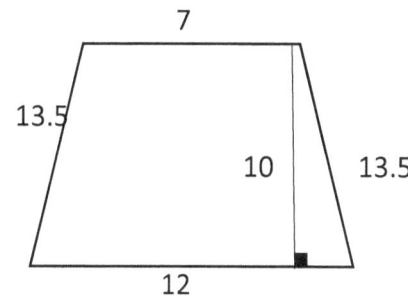

Perimeter: ....:

Area: ....:

5)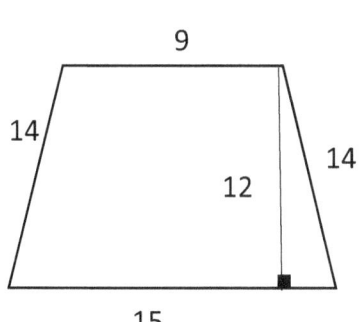

Perimeter: ....:

Area: ....:

6)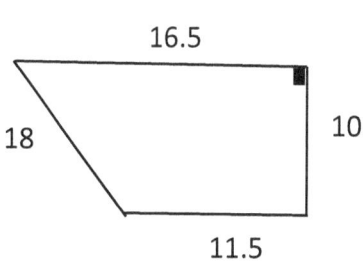

Perimeter: ....:

Area: ....:

Area and Perimeter of Parallelogram

Find the perimeter and area of each parallelogram.

1)

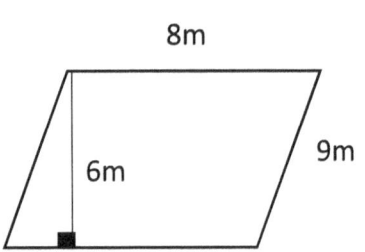

Perimeter: ....
Area: ....

2)

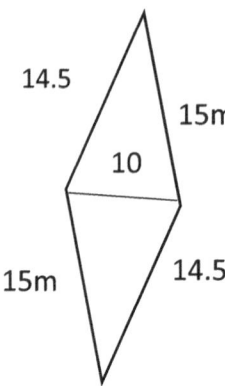

Perimeter: ....
Area: ....

3)

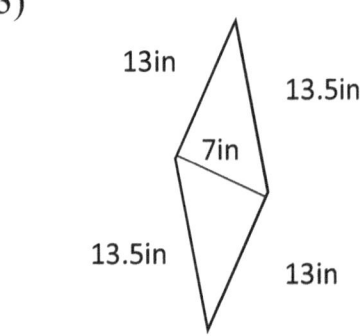

Perimeter: ....
Area: ....

4)

Perimeter: ....
Area: ....

5)

Perimeter: ....
Area: ....

6)

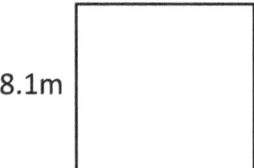

Perimeter: ....
Area: ....

Circumference and Area of Circle

Find the circumference and area of each ($\pi = 3.14$).

1)

Circumference:

Area:

2)

Circumference:

Area:

3)

Circumference:

Area:

4)

Circumference:

Area:

5)

Circumference:

Area:

6)

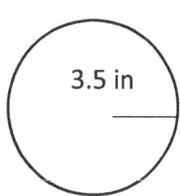

Circumference:

Area:

Perimeter of Polygon

Find the perimeter of each polygon.

1)
9.5mm

Perimeter: _____.

2)
4.5 m

Perimeter: _____:

3)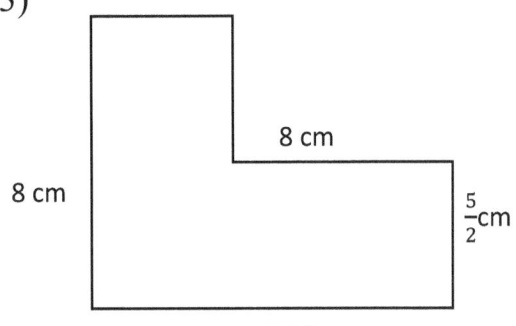
8 cm
8 cm
$\frac{5}{2}$ cm
12.5 cm

Perimeter: _____.

4)
6.2 in

Perimeter: _____:

5)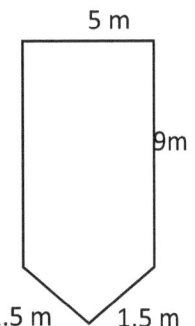
5 m
9m
1.5 m   1.5 m

Perimeter: _____.

6)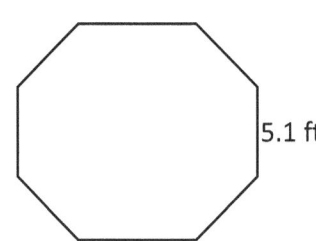
5.1 ft

Perimeter: _____:

Volume of Cubes

Find the volume of each cube.

1)

V:..................................

2)

V:..................................

3)

V:..................................

4)

V:..................................

5)

V:..................................

6)

V:..................................

Volume of Rectangle Prism

Find the volume of each rectangle prism

1)

V: ....

2)

V: ....

3)

V: ....

4)

V: ....

5)

V: ....

6)

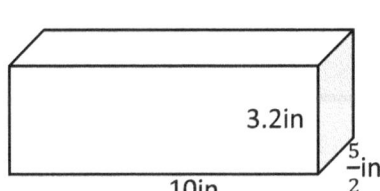

V: ....

Answer key Chapter 11

Area and Perimeter of Square

1. Perimeter: 16,   Area:16
2. Perimeter: 6,   Area:2.25
3. Perimeter: 10, Area:6.25
4. Perimeter: 32,   Area:64
5. Perimeter: 40,   Area:100
6. Perimeter: $4\sqrt{3}$, Area:3

Area and Perimeter of Rectangle

1- Perimeter: 16,   Area:15
2- Perimeter: 30,   Area:44
3- Perimeter: 22.4, Area:12
4- Perimeter: 23, Area:19
5- Perimeter: 24.4, Area:36
6- Perimeter:24, Area:30.24

Area and Perimeter of Triangle

1- Perimeter: 3s, Area:$\frac{1}{2}sh$
2- Perimeter: 22,   Area:12
3- Perimeter: 28.5, Area:42
4- Perimeter: 30,   Area:32
5- Perimeter: 38.5, Area:48
6- Perimeter: 60, Area:142.8

Area and Perimeter of Trapezoid

1- Perimeter: 49,   Area:100
2- Perimeter: 32,   Area:54
3- Perimeter: 38.5, Area:65
4- Perimeter: 46,   Area:95
5- Perimeter: 52,   Area:144
6- Perimeter: 56,   Area:140

Area and Perimeter of Parallelogram

1- Perimeter: $34m$, Area:$48(m)^2$
2- Perimeter: $59m$, Area:$145(m)^2$
3- Perimeter: $53in$, Area:$91(in)^2$
4- Perimeter: $30cm$, Area:$63(cm)^2$
5- Perimeter: $72.2m$,   Area:$295(m)^2$
6- Perimeter: $32.4m$, Area:$65.61(m)^2$

Circumference and Area of Circle

1) Circumference:37.68 mm   Area:$113.04(mm)^2$
2) Circumference: 7.536 in   Area:$4.522(in)^2$
3) Circumference: 15.7 m   Area:$19.625(m)^2$
4) Circumference: 69.08 cm   Area:$379.94(cm)^2$
5) Circumference: 25.12 km   Area:$50.24(km)^2$
6) Circumference: 21.98 in   Area:$38.465(in)^2$

Perimeter of Polygon

1) 47.5 mm
2) 27 m
3) 41 cm
4) 43.4 in
5) 26 m
6) 40.8 ft

Volume of Cubes

1) $125 m^3$
2) $343(mm)^3$
3) $729 in^3$
4) $2.197(cm)^3$

5) $42.875 (ft)^3$ 6) $9.261 (cm)^3$

Volume of Rectangle Prism

1) $288 (cm)^3$ 3) $38.75 (m)^3$ 5) $32.76 (mm)^3$
2) $882 (yd)^3$ 4) $231 (in)^3$ 6) $80 (in)^3$

Chapter 12: Data and Graphs

Mean and Median

Find the mean and median of the following data.

1) 12, 49, 23, 48, 15, 29, 69

Mean: __, Median: __

2) 8, 16, 16, 29, 17, 40

Mean: __, Median: __

3) 23, 75, 55, 18, 71, 19, 23, 10, 8

Mean: __, Median: __

4) 50, 22, 2, 30, 50, 36

Mean: __, Median: __

5) 9, 11, 19, 22, 43, 27, 4, 2, 44, 7

Mean: __, Median: __

6) 5, 13, 17, 35, 10, 33, 1, 66

Mean: __, Median: __

7) 20, 12, 33, 6, 19, 81, 9, 34, 25, 55

Mean: __, Median: __

8) 22, 17, 46, 13, 22

Mean: __, Median: __

9) 32, 32, 51, 23, 69

Mean: __, Median: __

10) 22, 11, 7, 8, 10, 12, 14

Mean: __, Median: __

11) 12, 36, 12, 18, 3, 24

Mean: __, Median: __

12) 16, 34, 28, 28, 11, 12, 11

Mean: __, Median: __

13) 5, 5, 6, 10, 2, 11, 30

Mean: __, Median: __

14) 33, 11, 15, 22, 22, 18, 39

Mean: __, Median: __

15) 12, 23, 23, 18, 34

Mean: __, Median: __

16) 7, 7, 19, 14, 16, 28, 31

Mean: __, Median: __

17) 31, 10, 12, 28, 19, 17, 6, 2

Mean: __, Median: __

18) 35, 39, 24, 58, 65, 4

Mean: __, Median: __

19) 43, 38, 19, 58, 3, 7

Mean: __, Median: __

20) 57, 46, 37, 18, 12, 66, 14

Mean: __, Median: __

Mode and Range

Find the mode(s) and range of the following data.

1) 16, 73, 10, 46, 12, 16, 57

Mode: __, Range: __

2) 7, 18, 18, 21, 14, 16

Mode: __, Range: __

3) 25, 32, 28, 49, 27, 25, 18, 13

Mode: __, Range: __

4) 28, 19, 3, 14, 37, 14

Mode: __, Range: __

5) 31, 18, 18, 15, 43, 15, 18, 59

Mode: __, Range: __

6) 8, 16, 12, 19, 8, 12, 24, 12

Mode: __, Range: __

7) 34, 18, 55, 46, 41, 34, 18, 34, 14, 34

Mode: __, Range: __

8) 72, 34, 7, 72, 29, 51

Mode: __, Range: __

9) 18, 18, 38, 52, 49

Mode: __, Range: __

10) 16, 10, 1, 1, 6, 12, 19

Mode: __, Range: __

11) 8, 17, 11, 9, 4, 9

Mode: __, Range: __

12) 16, 8, 22, 19, 21, 21, 42

Mode: __, Range: __

13) 5, 5, 2, 17, 4, 5, 29

Mode: __, Range: __

14) 14, 16, 20, 14, 14, 17, 27, 31

Mode: __, Range: __

15) 7, 16, 14, 16, 35

Mode: __, Range: __

16) 8, 8, 12, 19, 39, 43, 48

Mode: __, Range: __

17) 31, 19, 12, 33, 17, 21, 19, 1

Mode: __, Range: __

18) 48, 51, 48, 27, 67, 6

Mode: __, Range: __

19) 43, 15, 15, 72, 51, 3

Mode: __, Range: __

20) 40, 67, 28, 27, 14, 49, 67

Mode: __, Range: __

Stem–And–Leaf Plot

Make stem-and-leaf plots for the given data.

1) 25, 26, 47, 40, 22, 65, 28, 46, 48, 43, 29, 41, 66

   Stem | leaf

2) 51, 54, 19, 30, 52, 10, 36, 59, 38, 35, 39, 32

   Stem | leaf

3) 114, 88, 45, 45, 85, 117, 41, 116, 49, 112, 46, 89

   Stem | leaf

4) 71, 55, 76, 70, 106, 72, 101, 103, 79, 72, 59, 51

   Stem | leaf

5) 63, 68, 137, 64, 27, 139, 133

   Stem | leaf

6) 141, 37, 95, 32, 141, 148, 92, 146, 37, 97, 90

   Stem | leaf

Dot plots

The ages of students in a Math class are given below.

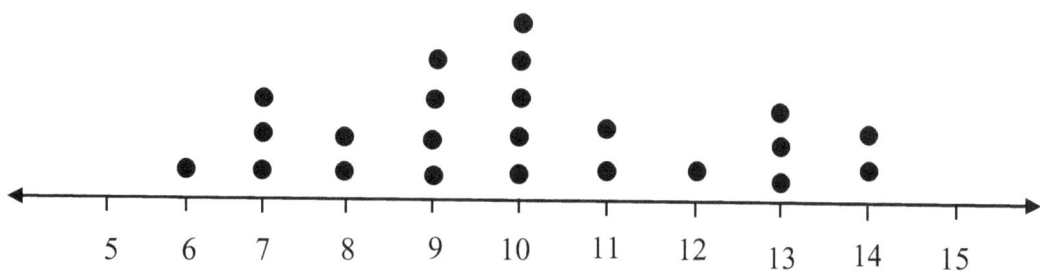

1) What is the total number of students in math class?

2) How many students are at least 13 years old?

3) Which age(s) has the most students?

4) Which age(s) has the fewest student?

5) Determine the median of the data.

6) Determine the range of the data.

7) Determine the mode of the data.

Common Core Math Practice Grade 5

Probability

1) A jar contains 15 caramels, 8 mints and 17 dark chocolates. What is the probability of selecting a mint?

2) If you were to roll the dice one time what is the probability it will NOT land on a 3?

3) A die has sides are numbered 1 to 6. If the cube is thrown once, what is the probability of rolling a 2?

4) The sides of number cube have the numbers 2, 4, 6, 2, 4, and 6. If the cube is thrown once, what is the probability of rolling a 4?

5) Your friend asks you to think of a number from one to ten. What is the probability that his number will be 7?

6) A person has 12 coins in their pocket. 3 dime, 5 pennies, 4 quarter, and a nickel. If a person randomly picks one coin out of their pocket. What would the probability be that they get a penny?

7) What is the probability of drawing an odd numbered card from a standard deck of shuffled cards (Ace is one)?

8) 40 students apply to go on a school trip. Three students are selected at random. what is the probability of selecting 5 students?

Bar Graph

Each student in class selected two games that they would like to play. Graph the given information as a bar graph and answer the questions below:

Game	Votes
Football	16
Volleyball	11
Basketball	19
Baseball	15
Tennis	11

1) Which was the most popular game to play?

2) How many more students like Basketball than Volleyball?

3) Which two game got the same number of votes?

4) How many Volleyball and Football did students vote in all?

5) Did more students like football or Volleyball?

6) Which game(s) did the fewest students like?

Answer key Chapter 12

Mean and Median

1) mean: 35, median: 29
2) mean: 21, median: 16.5
3) mean: 33.56, median: 23
4) mean: 31.67, median: 33
5) mean: 18.8, median: 15
6) mean: 22.5, median: 15
7) mean: 29.4, median: 22.5
8) mean: 24, median: 22
9) mean: 41.4, median: 32
10) mean: 12, median: 11
11) mean: 17.5, median: 15
12) mean: 20, median: 16
13) mean: 9.86, median: 6
14) mean: 22.86, median: 22
15) mean: 22, median: 23
16) mean: 17.43, median: 16
17) mean: 15.63, median: 14.5
18) mean: 37.5, median: 37
19) mean: 28, median: 28.5
20) mean: 35.71, median: 37

Mode and Range

1) mode: 16, range: 63
2) mode: 18, range: 14
3) mode: 25, range: 36
4) mode: 14, range: 34
5) mode: 18, range: 44
6) mode: 12, range: 16
7) mode: 34, range: 41
8) mode: 72, range: 65
9) mode: 18, range: 34
10) mode: 1, range: 18
11) mode: 9, range: 13
12) mode: 21, range: 34
13) mode: 5, range: 27
14) mode: 14, range: 17
15) mode: 16, range: 28
16) mode: 8, range: 40
17) mode: 19, range: 32
18) mode: 48, range: 61
19) mode: 15, range: 69
20) mode: 67, range: 53

Stem-And-Leaf Plot

1)

Stem	leaf
2	2 5 6 8 9
4	0 1 3 6 7 8
6	5 6

2)

Stem	leaf
1	0 9
3	0 2 5 6 8 9
5	1 2 4 9

3)

Stem	leaf
4	1 5 5 6 9
8	5 8 9
11	2 4 6 7

Common Core Math Practice Grade 5

4)

Stem	leaf
5	1 9 5
7	0 1 2 2 6 9
10	1 3 6

5)

Stem	leaf
2	7
6	3 4 8
13	3 7 9

6)

Stem	leaf
3	2 7 7
9	0 2 5 7
14	1 1 6 8

Dot plots

1) 23
2) 5
3) 10
4) 6 and 12
5) 2
6) 4
7) 2

Bar Graph

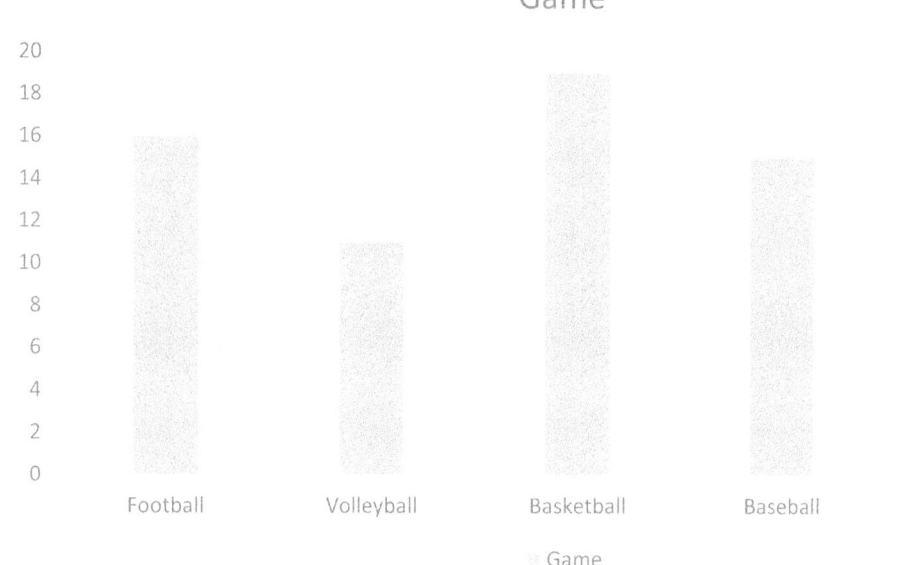

1) Basketball
2) 8 students
3) Volleyball and Tennis
4) 27
5) Football
6) Volleyball and Tennis

Probability

1) $\frac{1}{5}$
2) $\frac{5}{6}$
3) $\frac{1}{6}$
4) $\frac{1}{3}$
5) $\frac{1}{10}$
6) $\frac{5}{12}$
7) $\frac{5}{13}$
8) $\frac{1}{8}$

Common Core Test Review

Common Core GRADE 5 MAHEMATICS REFRENCE MATERIALS

Perimeter

Square $\qquad$ $P = 4S$

Rectangle $\qquad$ $P = 2L + 2W$

Area

Square $\qquad$ $A = S \times S$

Rectangle $\qquad$ $A = l \times w$ or $A = bh$

Volume

Cube $\qquad$ $A = S \times S \times S$

Rectangular Prism $\qquad$ $A = l \times w \times h$ or $A = Bh$

LENGTH

Customary	Metric
1 mile (mi) = 1,760 yards (yd)	1 kilometer (km) = 1,000 meters (m)
1 yard (yd) = 3 feet (ft)	1 meter (m) = 100 centimeters (cm)
1 foot (ft) = 12 inches (in.)	1 centimeter (cm) = 10 millimeters (mm)

VOLUME AND CAPACITY

Customary	Metric
1 gallon (gal) = 4 quarts (qt)	1 liter (L) = 1,000 milliliters (mL)
1 quart (qt) = 2 pints (pt.)	
1 pint (pt.) = 2 cups (c)	
1 cup (c) = 8 fluid ounces (Fl oz)	

WEIGHT AND MASS

Customary	Metric
1 ton (T) = 2,000 pounds (lb.)	1 kilogram (kg) = 1,000 grams (g)
1 pound (lb.) = 16 ounces (oz)	1 gram (g) = 1,000 milligrams (mg)

Common Core Practice Test 1

Mathematics

GRADE 5

Administered *Month Year*

Common Core Math Practice Grade 5

1) Which expression correctly shows the sum of product of 6 and 14 and the difference of 21 and 5?

   A. $14 + (6 \times 21) - 5$

   B. $(14 \times 6) + (21 - 5)$

   C. $(14 \times 6) - (21 - 5)$

   D. $14 - (6 \times 21) + 5$

2) Which equation has the same unknown value as $576 \div 32 = \square$?

   A. $576 \times \square = 32$

   B. $\square \div 576 = 32$

   C. $32 \times \square = 576$

   D. $\square \div 32 = 576$

3) The owner of a snow-cone stands used $\frac{1}{5}$ gallon of syrup to make 22 cherry snow cones. She used the same amount of syrup in each snow cone. How much syrup in gallons was used in each cherry snow cone?

   A. $\frac{1}{5}$ gal

   B. 5 gal

   C. $\frac{1}{110}$ gal

   D. 110 gal

4) 28 students equally share a bag of 588 dimes and 910 nickels. How many dimes does each student get?

   A. 21

   B. 22

   C. 28

   D. 53.5

5) Bertha bought 7 cans of tuna at $2.13 a can. How much did she spend?

   A. $9.67

   B. $14.91

   C. $14.63

   D. $13.91

6) Which equation shows how to multiply $18 \times 6 \times 7$ using the associative property?

   A. $18 \times 6 \times 7 = 7 + 6 \times 18$

   B. $(18 \times 6) + (18 \times 7) = (18 \times 7) + (18 \times 6)$

   C. $(18 \times 6) + 7 = 18 \times (6 + 7)$

   D. $(18 \times 6) \times 7 = 18 \times (6 \times 7)$

7) What is 0.63 rounded to the tenths place?

   A. 0.7

   B. 0.6

   C. 0.8

   D. 0.5

Common Core Math Practice Grade 5

8) Each time Elijah goes to the movies he spends $18.00. Which expression shows how much he spends after going to the movies t times?

   A. $18.00 + t

   B. $18.00 − t

   C. $18.00 ÷ t

   D. $18.00 × t

9) Benjamin is using a calculator to multiply 3,825 and 30. He enters 3,825 × 300 by mistake. What can Benjamin do to correct his mistake?

   A. Subtract 825 from the product

   B. Add 825 to product

   C. Divide the product by 10

   D. Multiply the product by 10

10) The price of shoes in a store is $24 and the price of belt in the same store is $15. A customer buys 3 shoes and 4 belts during a sale when the price of shoes is discounted 20% and the price of belt is discounted 6%. How much did the customer save due to the sale?

   A. $18

   B. $8.6

   C. $14.5

   D. $33.6

11) Daniel wrote the expression shown. $32 \div 8 + 7(85 - 10)$. What do these parentheses indicate in the expression?

   A. Divide 32 by 8 before adding 7

   B. Multiply 7 by 85 before subtracting 10

   C. Subtract 10 from 85 before multiplying by 7

   D. Add 8 and 7 together before subtracting 10 from 85

12) Lucas is doing his homework. It takes him 9 minutes to do 2 problems of math and 40 minutes to read 16 pages of biology. Lucas reads the same amount of biology pages each minute. Which of these is closest to the number of minutes it takes Lucas to read each page of his biology homework?

   A. 4 minutes

   B. $2\frac{1}{2}$ minutes

   C. 4 minutes

   D. $1\frac{1}{2}$ minutes

13) What is the sum of angles α + β in the right triangle below?

   A. 90 degrees

   B. 45 degrees

   C. 135 degrees

   D. 180 degrees

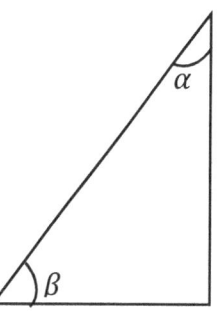

14) Which statement describe $\frac{2}{9} \times \frac{5}{12}$?

  A. $\frac{2}{9} \times \frac{5}{12}$ is 5 group of $\frac{2}{9}$, divided into 108 equal parts.

  B. $\frac{2}{9} \times \frac{5}{12}$ is 5 group of $\frac{2}{9}$, divided into 12 equal parts.

  C. $\frac{2}{9} \times \frac{5}{12}$ is 12 group of $\frac{2}{9}$, divided into 5 equal parts.

  D. $\frac{2}{9} \times \frac{5}{12}$ is 12 group of $\frac{2}{9}$, divided into 10 equal parts.

15) Noah and Liam played a new game for 14 hours last week. If they played the same amount of time each of 8 days, how long did they play each day?

  A. 0.5 hour

  B. 0.25 hour

  C. 1.5 hour

  D. 1.75 hour

16) Daniel planted five cucumber seeds. Out of the five planted, only three sprouted. How many plants can Daniel plan on yielding if he plants 200 seeds?

Number of seeds	6				300
Successes	5				?

  A. 120

  B. 50

  C. 150

  D. 250

17) A wire is 30 feet long. The Robert needs to cut pieces that are $\frac{3}{7}$ foot long. How many pieces can he cut?

   A. 70

   B. 17

   C. 7

   D. 35

18) The temperature was 43°F at 7:00 A.M., 49°F at 7:30 A.M., and 55°F at 8:00 A.M. Describe the pattern, and predict the temperature at 09:30 A.M.

   A. Add 5°F; 75°F

   B. Add 5°F; 73°F

   C. Add 6°F; 70°F

   D. Add 6°F; 73°F

19) A rectangular prism measures 12-unit cubes wide and 3-unit cubes high. If the volume of the prism is 180 cubic unit, what is the length of the prism in unit cubes?

   A. 5

   B. 15

   C. 36

   D. 60

Common Core Math Practice Grade 5

20) Evelyn and Mason are both saving to buy cars. So far, Mason has saved $2,352. Evelyn has saved 5 times as much as Mason. How much has Evelyn saved?

   A. $10,760

   B. $11,670

   C. $11,760

   D. $12,670

21) Which equation is true when m = 12?

   A. $12 + 24 \div m = 6$

   B. $24 \div m + 12 = 3$

   C. $36 + 12 \div m = 10$

   D. $24 \div m + 4 = 6$

22) Which is equivalent to 33 meters?

   A. 0.33 cm

   B. 3.3 cm

   C. 330 cm

   D. 3,300 cm

23) If k is an odd integer, which of the following must be an even integer?

   A. $k + 1$

   B. $2k - 1$

   C. $2k + 1$

   D. $k + 2$

24) Charlotte has a grosgrain ribbon 7 feet long. She cuts the ribbon into 10 equal pieces. Which equation shows how to find the length, in feet of each piece of the ribbon?

   A. $10 \times 7 = 70$

   B. $7 \div 10 = \frac{7}{10}$

   C. $10 + 7 = 17$

   D. $7 \div 10 = 1\frac{2}{7}$

25) Paul and his five friends had lunch together. The total bill for lunch came to $42.60, including tip. If they shared the bill equally, how much did they each pay?

   A. 6.10

   B. 8.52

   C. $7.10

   D. 8.10

26) What is the volume of the figure in cubic units?

   A. 182

   B. 222

   C. 150

   D. 244

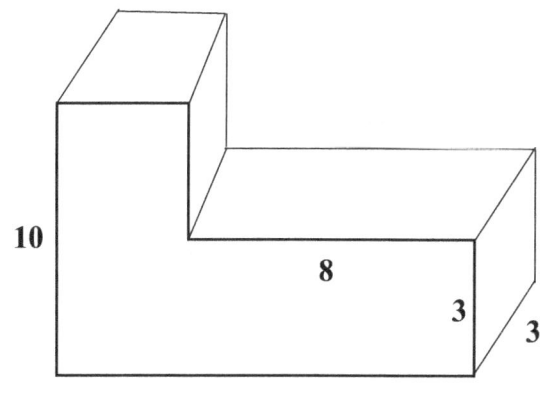

Common Core Math Practice Grade 5

27) The graph shows three of four vertices of parallelogram ABCD. At which location on the coordinate grid could point D be located?

A. (1.5, 2)

B. (1.5, 2)

C. (2, 5.5)

D. (0.5, 2)

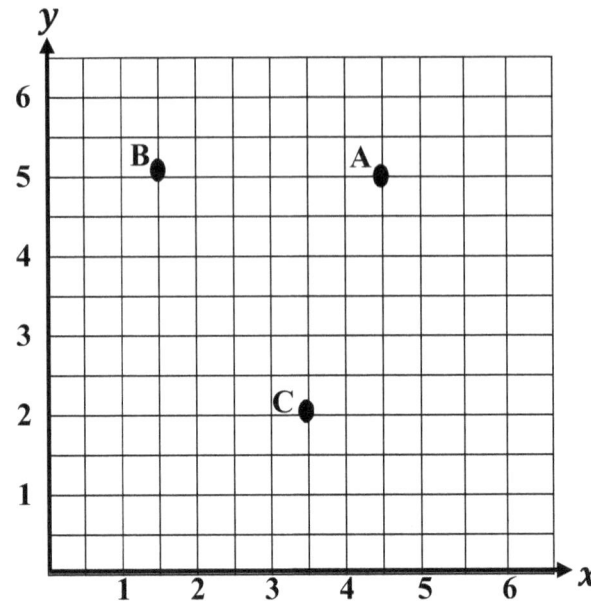

28) Which is a composite number?

A. 17

B. 47

C. 87

D. 97

29) Which capacities are written in order from greatest to least?

A. 6,217 mL; 6.14 L; 614 mL; 7 L; 8,002 mL

B. 8,002 mL; 7 L; 6,217 mL; 6.14 L; 614 mL

C. 614 mL; 6.14 L; 6,217 mL; 7 L; 8,002 mL

D. 7L; 8,002 mL; 6.14 L; 6,217 mL; 614 mL

30) Mrs. Smith's family ordered 10 pizzas. They ate $5\frac{1}{4}$ pizzas and gave $3\frac{7}{8}$ pizzas to her friend. How many pizzas do they have left?

A. $\frac{7}{8}$

B. $1\frac{1}{8}$

C. $1\frac{7}{8}$

D. $9\frac{1}{8}$

31) There is a big carnival every year in our town. It costs $8.00 to get in and $2.20 for every ride ticket you buy, but some of them take four-ticket rides. How much does it cost to get in and buy 10 four-ticket rides?

A. $58.80

B. $40

C. $88

D. $96

32) A cake pan, and its side lengths are shown. What does the perimeter of the cake pan in inches?

A. 27.2 in.

B. 17.8 in.

C. 27 in.

D. 17.2 in.

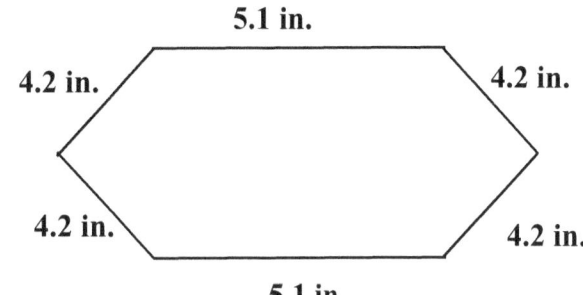

33) Eugen has a piggy bank filled with coins.

- 11 are quarters
- 22 are dimes
- 20 are nickels
- 47 are pennies

What percent of the coin are nickels?

A. 20%

B. 45%

C. 55%

D. 100%

34) Sophia is buying crullers for a group of 35 students. If 60% of the group want crullers, and each student will eat 3 crullers, how many crullers should Sophia buy?

A. 12 crullers

B. 21 crullers

C. 63 crullers

D. 36 crullers

35) Which fraction is equivalent to $\frac{5}{9}$?

A. $\frac{45}{80}$

B. $\frac{10}{27}$

C. $\frac{30}{63}$

D. $\frac{35}{63}$

Common Core Math Practice Grade 5

36) Jayden is having a party. He wants to get three cookies for each of the 5 people, including himself, who will be at the party. If each cookie costs 60 ₵, how much money will he spend on cookies? which diagram below best shows this problem?

A.   B.   C.   D.

37) What is the perimeter of the rectangular in the figure below?

A. 7.68

B. 12.8

C. 11

D. 11.2

3.2 in, 2.4 in

38) What is the value of C in the table below?

A. 151

B. 130

C. 150

D. 152

Input	Output
2	3
6	15
14	39
30	87
51	C

39) Robert went fishing and caught 3 fish. They weighed 17.41 ounces, 33.74 ounces, and 89.32 ounces. How much did all 3 Robert's fishes weigh together?

A. 138.73 ounces

B. 168.52 ounces

C. 140.47 ounces

D. 137.64 ounces

40) The bar graph shows the monthly high sale for a grocery store in 2018. According to the graph, for all months shown, how much smaller is the mean than the median sale?

A. 0 Million Dollars

B. 4 Million Dollars

C. 22 Million Dollars

D. 35 Million Dollars

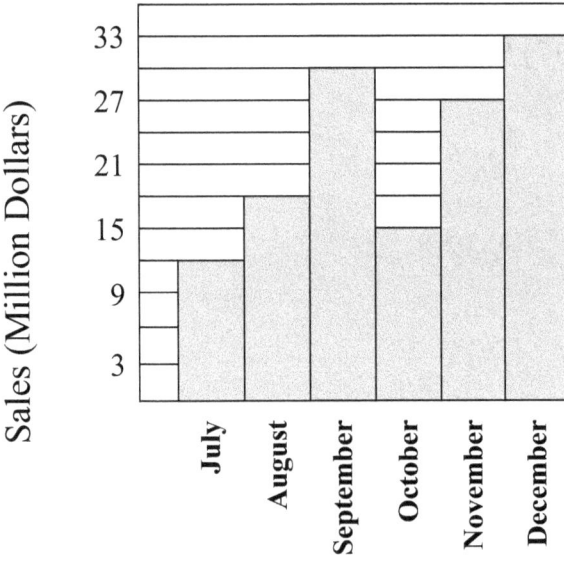

Common Core Practice Test 2

Mathematics

GRADE 5

Administered Month Year

Common Core Math Practice Grade 5

1) Which digit is the hundreds digit in the number 79,530,214?

    A. 4

    B. 1

    C. 2

    D. 9

2) What is the value of expression shown? $3[6.3 - 4(1.1)]$

    A. 5.7

    B. 7.59

    C. 5.75

    D. 14.5

3) Which expression equal to $\frac{54}{90}$?

    A. $\frac{4}{5}$

    B. $\frac{9}{11}$

    C. $\frac{3}{7}$

    D. $\frac{3}{5}$

4) How can the distributive property be used to solve this expression? 72×36

    A. $(70 + 30) \times (2 \div 6)$

    B. $(7 \times 3) + (2 \times 6)$

    C. $(72 + 3) \times (72 + 6)$

    D. $(72 \times 30) + (72 \times 6)$

5) Which statement correctly compares the two values?

   A. The value of 7 in 7.59 is 10 times the value of the 7 in 9.75

   B. The value of 7 in 9.75 is $\frac{1}{10}$ the value of the 7 in 7.59

   C. The value of 7 in 9.75 is 100 times the value of the 7 in 7.59

   D. The value of 7 in 9.75 is $\frac{1}{100}$ the value of the 7 in 7.59

6) What is 83.61 ÷ 100?

   A. 8.361

   B. 0.8361

   C. 83.61

   D. 836.1

7) Which number represents six million fifty thousand three hundred four?

   A. 6,050,034

   B. 6,500,304

   C. 6,050,304

   D. 6,500,034

8) Betty has 0.7 liters of juice. How many milliliters(mL) of juice does Betty have?

   A. 0.007 mL

   B. 0.0007 mL

   C. 7,000 mL

   D. 700 mL

9) What is sum of 8.75 and 3.7?

   A. $11\frac{29}{200}$

   B. $11\frac{73}{100}$

   C. $12\frac{1}{45}$

   D. $12\frac{9}{20}$

10) Joshua road her bicycle 21.6 miles in 3 hours. Which is how far he road in 8 minutes?

    A. 0.56 mile

    B. 0.96 miles

    C. 7.20 mile

    D. 0.12 miles

11) Mr. Johnson paid $34.56 for each adult shirt and $21.45 for each youth shirt he bought. Mr. Johnson bought 3 adult shirts and 4 youth shirts. How much money did he spend on these shirts?

    A. $108.68

    B. $158.85

    C. $189.48

    D. $282.48

Common Core Math Practice Grade 5

12) Mr. Martinez had 16 daylily plants. Each plant produced 798 flowers. How many flowers did the plants produce?

   A. 98

   B. 1,278

   C. 2,778

   D. 12,768

13) The table below shows the lengths of different Stamps on display at a post office. Which stamps has the shortest length?

   A. Stamp 1

   B. Stamp 2

   C. Stamp 3

   D. Stamp 4

Stamp	Length (cm)
1	$\frac{2}{3}$
2	$\frac{5}{6}$
3	$\frac{7}{18}$
4	$\frac{5}{9}$

14) Ms. Bianchi works 24.5 hours a week at the movie theater. She earns $14 an hour. Which statement about her weekly income is true?

   A. Her gross income is less than $343.

   B. Her net income is less than $343.

   C. Her gross income is more than $343.

   D. Her net income is more than $343.

Common Core Math Practice Grade 5

15) A market charges $1.60 for a dozen of eggs. How many dozen eggs can you buy with $6.40?

   A. 5

   B. 4

   C. 6

   D. 8

16) An art teacher has $900. She spent $183.49 on art supplies and $512.34 on new desks. How much money does she have left?

   A. $204.17

   B. $240.70

   C. 148.17

   D. $695.83

17) Find the value of the nth term in the sequence.

Position	10	12	14	16	n
Value of Terms	3	4	5	6	

   A. $\frac{n}{5}$

   B. $n-8$

   C. $\frac{n}{2} - 2$

   D. $\frac{n}{2} + 3$

18) The dot plot shows the number of students who made from 1 to 8 words. What fraction of the student in the class made 6 or more words?

A. $\frac{1}{4}$

B. $\frac{1}{3}$

C. $\frac{5}{6}$

D. $\frac{2}{3}$

19) Jessica flips a coin 60 times. The coin lands on head 15 times. What percent of the flips were tails?

A. 25%

B. 15%

C. 85%

D. 75%

20) A parallelogram has a base of 31 cm and the height of 16 cm. a triangle has a base of 54 cm and a height of 18 cm. which figure has the greater area? By how much? (*Area of parallelogram = base × high*)

A. Parallelogram, 10 cm^2

B. Parallelogram, 15 cm^2

C. Triangle, 10 cm^2

D. Triangle, 15 cm^2

Common Core Math Practice Grade 5

21) Which of the following fractions is higher than $\frac{1}{6}$, but less than $\frac{1}{3}$?

   A. $\frac{3}{4}$

   B. $\frac{4}{5}$

   C. $\frac{1}{8}$

   D. $\frac{1}{5}$

22) $\frac{17}{20}$ may be written as a percent as:

   A. 15%

   B. 25%

   C. 75%

   D. 85%

23) A chart below is showing the socks that Jenna has in her dresser. What is the probability that Jenna will choose a red sock at random?

   A. $\frac{1}{2}$

   B. $\frac{1}{3}$

   C. $\frac{1}{7}$

   D. $\frac{1}{5}$

Color	Number
Purple	6
White	1
Red	7
Stripped	2
Black	5

24) The measures of three of the interior angles of a quadrilateral are 70°, 120°, and 85°. What is the measure of the fourth angle of this quadrilateral?

A. 35°

B. 85°

C. 95°

D. 170°

25) What is the greatest common factor (GCF) of 64 and 36?

A. 2

B. 3

C. 4

D. 16

26) Which inequality is true?

A. $0.89 > 0.98$

B. $0.55 < 0.54$

C. $0.35 > 0.53$

D. $0.46 < 0.64$

27) Which percent is equivalent to 0.6?

A. 6%

B. 0.6%

C. 0.006%

D. 60%

Common Core Math Practice Grade 5

28) James paid $11.40 each for 4 used video games and $24.5 each for 3 new video games. How much money did James spend on video games?

   A. $119.1

   B. $19.10

   C. $101.9

   D. $191.1

29) The arcade changes the price shown below for game tokens. Using the pattern shown, what is the price for 8 tokens?

   A. $2.40

   B. $4.20

   C. $1.20

   D. $2.10

Tokens	Amount
10	$4.00
20	$8.00
30	$12.00
40	$16.00
50	$20.00

30) Suzy's class used $\frac{7}{9}$ of the cafeteria trays for their science display. There are 63 trays in all. How many trays were used?

   A. 7

   B. 36

   C. 49

   D. 630

31) There are 1.4124 kilograms in a pound. How many kilograms are in 3 pounds?

   A. 4.3227

   B. 4.2372

   C. 4.7232

   D. 4.2327

32) Which of the following angles is acute?

   A. 60 Degrees

   B. 110 Degrees

   C. 95 Degrees

   D. 120 Degrees

33) A water tank contains 795 liters of water. A leak in the tank causes 1.25 liters to drip out each hour. How much water will the tank contain after 3 days?

   A. 72 liters

   B. 795 liters

   C. 705 liters

   D. 772 liters

34) If a secretory paid $58.50 for 9 rolls of film, how much did she pay for each roll?

   A. $6. 50

   B. $3.50

   C. $65.50

   D. $651.50

35) The perimeter of the trapezoid below is 44. What is its area?

A. 44 cm

B. 108 cm

C. 118 cm

D. 81.5 cm

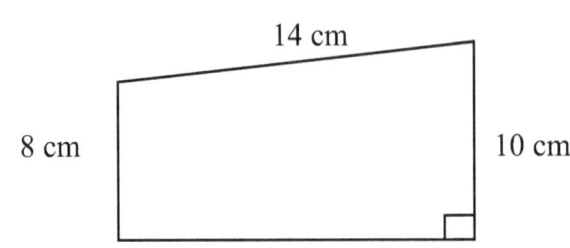

36) Joseph's first planter is 8 feet long and 3 feet wide and 4 feet height. The container is filled with soil to a height of 3 feet in the planter. What is the volume of soil in the planter?

A. 69 cubic cm

B. 66 cubic cm

C. 84 cubic cm

D. 96 cubic cm

37) The stem and leaf plot shows the numbers of miles run by each of runners. What is the difference between the number of runners who ran fewer than 40 miles and the number of runners who ran more than 75 miles?

A. 17

B. 14

C. 6

D. 3

Stem	Leaf
0	8
1	2 4 7 9
2	0 1 1 7 9
4	2 4 8 8
6	0 3 5
8	4 6 9
9	1

38) Linda will order ice-cream for her party. The relationship between S, the number of the scoops she will order, and C, the total costs of ice-cream she will pay, can be represented by the equation C = 8S.

Which table contain only values that represent the equation?

A. **Invoice**

Number of Scoops (S)	Cost (C)
4	$32
8	$64
11	$88
14	$108

B. **Invoice**

Number of Scoops (S)	Cost (C)
4	$32
8	$64
11	$88
14	$112

C. **Invoice**

Number of Scoops (S)	Cost (C)
2	$18
6	$48
10	$80
13	$104

D. **Invoice**

Number of Scoops (S)	Cost (C)
2	$16
6	$58
10	$80
13	$96

39) A cube has a volume of 27 cubic units. How many unit cubes will fit along one side of the cube if there are no gaps on overlaps?

A. 3

B. 6

C. 9

D. 27

40) Figure M and figure N are congruent. What is the length of side b in figure N?

A. 8

B. 4

C. 5

D. 6

Figure M

Figure N

Answers and Explanations

Answer Key

Now, it's time to review your results to see where you went wrong and what areas you need to improve!

Common Core Math Practice Tests

Practice Test 1

1	B	16	D	31	D	
2	C	17	A	32	C	
3	C	18	D	33	A	
4	A	19	A	34	C	
5	B	20	C	35	D	
6	D	21	D	36	D	
7	B	22	D	37	D	
8	D	23	A	38	C	
9	C	24	B	39	C	
10	A	25	C	40	A	
11	C	26	B			
12	B	27	D			
13	A	28	C			
14	B	29	B			
15	D	30	A			

Practice Test 2

1	C	16	A	31	B	
2	A	17	C	32	A	
3	D	18	B	33	C	
4	D	19	D	34	A	
5	A	20	A	35	B	
6	B	21	D	36	D	
7	C	22	D	37	C	
8	D	23	B	38	B	
9	D	24	B	39	A	
10	B	25	C	40	C	
11	C	26	D			
12	D	27	D			
13	C	28	A			
14	B	29	A			
15	B	30	C			

Practice Test 1

Answers and Explanations

1) **Answer: B**

   The product of 6 and 14 is (6×14)

   The difference of 21 and 5: $(21 - 5)$

   Then, the sum: $(6 \times 14) + (21 - 5)$

2) **Answer: C**

   Multiplication is the inverse of division and you can use multiplication to check your division answer (Dividend÷Divisor =Quotient; Then, Quotient × Divisor = Dividend).

   $576 \div 32 = \Box \rightarrow \Box \times 32 = 576$

3) **Answer: C**

   $\frac{1}{5} \div 22 = \frac{1}{5} \times \frac{1}{22} = \frac{1}{110}$

4) **Answer: A**

   The bag of dimes should be divided by all students: $588 \div 28 = 21$ dimes

5) **Answer: B**

   $7 \times 2.13 = \$14.91$

6) **Answer: D**

   Associative property of multiplication: when multiplying three or more real numbers, the product is always the same regardless of their regrouping. By 'grouping' we mean how you use parenthesis. $(a \times b) \times c = a \times (b \times c)$

7) **Answer: B**

   To round a number to the nearest tenth, look at the next place value to the right (the hundredths). If it is 4 or less, just remove all the digits to the right. If it is 5 or greater, add 1 to the digit in the tenths place, and then remove all the digits to the right.

8) **Answer: D**

   Each time spend $18 and, t times, means multiplication, then we need to use multiplication. The answer is: 18t

9) **Answer: C**

$3{,}825 \times 300 = 3{,}825 \times 30 \times 10$, then $\frac{3{,}825 \times 30 \times 10}{10} = 3{,}825 \times 30$

10) **Answer: A**

3 Shoes: $3 \times 24 = 72$; 20% of 72: $0.20 \times 72 = \$14.4$

4 Belts: $4 \times 15 = 60$; 6% of 60: $0.06 \times 60 = \$3.6$

$\$14.4 + \$3.6 = \$18$ the customer saved

11) **Answer: C**

The **PEMDAS Rule** (an acronym for "Parenthesis, Exponents, Multiplication, Division, Addition, Subtraction") is a set of rules that prioritize the order of calculations, that is, which operation to perform first. Parenthesis in math are used to group important things together, so you always do them first.

12) **Answer: B**

$40 \div 16 = \frac{40}{16} = 2\frac{8}{16} = 2\frac{1}{2}$ minutes

13) **Answer: A**

The sum of the measure of the three interior angles, in every triangle, is 180°.

A right triangle is a type of triangle that has one angle that measures 90°, then:

$90° + \alpha + \beta = 180° \rightarrow \alpha + \beta = 180° - 90° = 90°$

14) **Answer: B**

To multiply fractions is following the 3 steps: Multiply the numerators, Multiply the denominators, and Simplify the resulting fraction.

$\frac{2}{9} \times \frac{5}{12} = \frac{5}{12} \times \frac{2}{9} = (\frac{1}{12} \times \frac{5}{1}) \times \frac{2}{9} = \frac{1}{12} \times (5 \times \frac{2}{9}) = \frac{5 \times \frac{2}{9}}{12}$

15) **Answer: D**

$14 \div 8 = 1.75$

16) **Answer: D**

Equivalent Fractions have the same value, even though they may look different. Because when you multiply or divide both the top and bottom by the same number, the fraction keeps its value. Then, $\frac{6}{5} = \frac{300}{?} \rightarrow \frac{6 \times 50}{5 \times 50} = \frac{300}{250}$

Common Core Math Practice Grade 5

17) **Answer: A**

30 feet long wire is cut into $\frac{3}{7}$ equal parts. Therefore, 30 should be divided by $\frac{3}{7}$

$30 \div \frac{3}{7} = 30 \times \frac{7}{3} = 70$

18) **Answer: D**

The pattern is: $43°, 49°, 55°$ (Add $6°$).

09:30A.M.$-$8:00A.M.=1:30 hours, and change to the half hours is $2 \times 1.5 = 3$, and use the pattern: $3 \times 6 = 18$; then, $18° + 55° = 73°$

19) **Answer: A**

Use volume of rectangle prism. V = width $\times$ length $\times$ heigth

$\Rightarrow 180 = 12 \times L \times 3 \Rightarrow 180 = 36L \Rightarrow L = \frac{180}{36} = 5$ uint-cubes

20) **Answer: C**

$2,352 \times 5 = 11,760$

21) **Answer: D**

Plug in 6 for m in the equations then, use order of operation rule (PEMDAS):

A. $12 + 24 \div m = 12 + 24 \div 12 = 12 + 2 = 14 \rightarrow 14 \neq 6$

B. $24 \div m + 12 = 24 \div 12 + 12 = 2 + 12 = 14 \rightarrow 14 \neq 3$

C. $36 + 12 \div m = 36 + 12 \div 12 = 36 + 1 = 37 \rightarrow 37 \neq 10$

D. $24 \div m + 4 = 24 \div 12 + 4 = 2 + 4 = 6 \rightarrow 6 = 6$

22) **Answer: D**

1 m=100 cm $\Rightarrow$ 33 m = 3,300 cm

23) **Answer: A**

We can use an odd number as examples. If k is odd $2k$ is even (k=3, 6), and $2k \pm 1$ is odd (k=3,7or 5), and $k + 2$, is odd (k=1, 3). then, if k is odd, $k + 1$ is even (k=1, 2).

24) **Answer: B**

7 feet long ribbon is cut into 10 equal parts. Therefore, 7 should be divided by 10

25) **Answer: C**

$42.60 \div 6 = \$7.1$

Common Core Math Practice Grade 5

26) Answer: B

We have two rectangular prisms.

Use volume of rectangle prism. V = width × length × heigth

$V_1 = 8 \times 3 \times 3 = 72$ cubic units

$V_2 = 10 \times 5 \times 3 = 150$ cubic units

$V = V_1 + V_2 = 72 + 150 = 222$ cubic unit

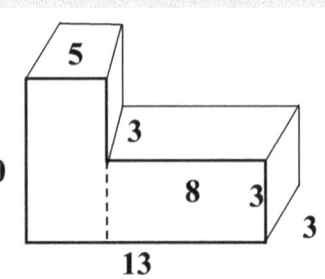

27) Answer: D

To determine the location of point D, the last vertex (corner) of the parallelogram (a four-sided figure with two pairs of parallel sides), you should have found a point exactly three units away (or 6 grid squares that are each 0.5 units long) from point C on the same horizontal grid line. Then, the ordered pairs that represents this location are (6.5, 2) by right side and (0.5, 2) by left side.

28) Answer: C

Composite numbers are a whole number that is a multiple of at least two numbers other than itself and 1. They have factors. $87 = 3 \times 29$.

29) Answer: B

First convert mL to L, then order from greatest:

614 mL=0.614 L;   6,217 mL=6.217 L,   8,002 mL=8.002 L

8.002 L, 7 L, 6.217 L, 6.14 L, 0.614 L

8,002 mL, 7 L, 6,217 mL, 6.14 L, 614 mL

30) Answer: A

$5\frac{1}{4} + 3\frac{7}{8} = 5\frac{2}{8} + 3\frac{7}{8} = 8\frac{2+7}{8} = 8\frac{9}{8} = 9\frac{1}{8}$

$10 - 9\frac{1}{8} = \frac{7}{8}$ Pizza

31) Answer: D

$10 \times 4 = 40$ tickets and each ticket cost 2.20: $40 \times \$2.20 = \$88$

Total cost: $\$88 + \$8 = \$96$

Common Core Math Practice Grade 5

32) Answer: C

To determine the perimeter (distance around the outside) of the hexagon, you should have added all of the side lengths.

$4.2 + 4.2 + 5.1 + 4.2 + 4.2 + 5.1 = 27$ in.

33) Answer: A

Percent means "out of 100." Or, Use percent formula: $percent = \frac{part}{whole} \times 100$

Whole$= 11 + 22 + 20 + 47 = 100$,

The nickels are 20, means: $\frac{20}{100}$ or 20%

34) Answer: C

Use percent formula: $part = \frac{percent}{100} \times whole$

Part$= \frac{60}{100} \times 35 = \frac{60 \times 35}{100} = \frac{2,100}{100} = 21$ students,

$21 \times 3 = 63$ crullers

35) Answer: D

Multiplying or dividing the numerator and denominator of a fraction by the same number will produce an equivalent fraction. $\frac{5}{9} \times \frac{7}{7} = \frac{35}{63}$

36) Answer: D

He wants to buy 3 cookies for 5 people: $3 \times 5 = 15$ cookies

Each cookie cost 60 cents, then, we need to have 15 of ¢60 or 15×60.

37) Answer: D

Perimeter is $P = 2l + 2w$, where l is the length and w is the width

$p = 2(3.2) + 2(2.4) = 6.4 + 4.8 = 11.2$ in.

38) Answer: C

We need to find the relation between input and output.

Input: $(6 - 2) = 4$, Output: $(15 - 3) = 12$, and the relation is 3 times of 4 is 12

Input, $(51 - 30) = 21$, use the relation: $3 \times 21 = 63$, then:

Output, $C - 87 = 63 \rightarrow C = 63 + 87 = 150$

39) Answer: C

$17.41 + 33.74 + 89.32 = 140.47$

40) Answer: A

The sales are: 12, 18, 30, 15, 27, and 33

Write the numbers in order: 12, 15, 18, 27, 30, 33

Median: $\frac{18+27}{2} = \frac{45}{2} = 22.5$

Mean: $\frac{\text{sum of terms}}{\text{number of terms}} = \frac{12+15+18+27+30+33}{6} = \frac{135}{6} = 22.5$

Median − Mean = $22.5 - 22.5 = 0$

Practice Test 2

Answers and Explanations

1) **Answer: C**

Place Value Chart								
Millions			Thousands			Ones		
Hundreds	Tens	Ones	Hundreds	Tens	Ones	Hundreds	Tens	Ones
	7	9	5	3	0	2	1	4

In the number 79,530,214 the digit 2 is in the (one) hundreds place.

2) **Answer: A**

You should have used the order of operations, or PEMDAS. 1. Operations contained in Parentheses or brackets, 2. Exponents 3. Multiplication/Division from left to right, and 4. Addition/Subtraction from left to right.

$3[6.3 - 4(1.1)] = 3[6.3 - 4.4] = 3[1.9] = 5.7$

3) **Answer: D**

Multiplying or dividing the numerator and denominator of a fraction by the same number will produce an equivalent fraction. $\frac{54}{90}$ divided by 6 = $\frac{9}{15}$ divided by 3 = $\frac{3}{5}$

4) **Answer: D**

The distributive property lets you multiply a sum by multiplying each addend separately and then add the products.

$72 \times 36 = 72 \times (30 + 6) = (72 \times 30) + (72 \times 6)$

5) **Answer: A**

The place value of 7 in 9.75 is tenth

The place value of 7 in 7.59 is ones

To convert tenth place to ones we need to multiply by 10.

6) **Answer: B**

$83.61 \div 100 = 0.8361$

Common Core Math Practice Grade 5

7) Answer: C

When converting word names to standard form. The words "million" and "thousand" tell you which periods the digits are in.

8) Answer: D

1 L = 1,000 mL; Convert 0.7 L to mL: $0.7 \times 1,000 = 700$ mL

9) Answer: D

Add decimals: $8.75 + 3.7 = 12.45$

Convert decimal to fraction: $12.45 = 12\frac{45}{100}$, then simplify fraction (by5): $12\frac{9}{20}$

10) Answer: B

$21.6 \div 3 = 7.2$ in one hour

$7.2 \div 60 = 0.12$ in one minute,

$8 \times 0.12 = 0.96$ mile

11) Answer: C

$3 \times 34.56 + 4 \times 21.45 = 103.68 + 85.8 = 189.48$

12) Answer: D

$16 \times 798 = 12,768$

13) Answer: C

Compare the fractions: $\frac{5}{6} > \frac{2}{3} > \frac{5}{9} > \frac{7}{18}$

Therefore, $\frac{7}{18}$ is the least fraction and related to stamp 3

14) Answer: B

You should have first determined that her weekly gross income (income before paying taxes) is $24.5 \times \$14 = \$343$. But, her weekly net income (income after paying taxes) would be less than her weekly gross income of $343.

15) Answer: B

$6.40 \div 1.60 = 4$

16) Answer: A

$183.49 + 512.34 = 695.83$, then subtract from $900: $900 - 695.83 = 204.17$

WWW.MathNotion.com

Common Core Math Practice Grade 5

17) Answer: C

Plug in $n = 10$ and $n = 12$ in each equation.

A. $\frac{n}{5} \rightarrow \frac{10}{5} = 2 \neq 3$

B. $-8 \rightarrow 10 - 8 \neq 4$

C. $\frac{n}{2} - 2 \rightarrow \frac{10}{2} - 2 = 5 - 2 = 3$ and $\frac{12}{2} - 2 = 6 - 2 = 4$ Bingo!

D. $\frac{n}{2} + 3 \rightarrow \frac{10}{2} + 3 = 5 + 2 \neq 4$

18) Answer: B

All student is 12, and 4 of them made 6 and more, then $\frac{4}{12}$ and simplify: $\frac{1}{3}$

19) Answer: D

Use percent formula: $percent = \frac{part}{hole} \times 100$

$Percent = \frac{15}{60} \times 100 = 25\%$ coin lands on head. 100%−25%=75% land on tail.

20) Answer: A

Area of parallelogram: $A = b.h = 31 \times 16 = 496 \ cm^2$

Area of triangle: $A = \frac{1}{2}b.h = \frac{1}{2} \times 54 \times 18 = 486 \ cm^2$

Area of triangle− Area of parallelogram $= 496 - 486 = 10 \ cm^2$

21) Answer: D

Compare the fractions:

$\frac{3}{4} > \frac{1}{3}; \quad \frac{4}{5} > \frac{1}{3}; \quad \frac{1}{8} < \frac{1}{6}; \quad \frac{1}{5} > \frac{1}{6}$, and $\frac{1}{5} < \frac{1}{3}$

22) Answer: D

Convert fraction to decimal: $\frac{17}{20} \times 100 = \frac{17 \times 100}{20}$ both numerator and denominator divided by 20: $\frac{17 \times 5}{1} = 85\%$

23) Answer: B

Probability $= \frac{number \ of \ desired \ outcoes}{number \ of \ total \ outcomes} = \frac{7}{6+1+7+2+5} = \frac{7}{21} = \frac{1}{3}$

WWW.MathNotion.com

24) Answer: B

Quadrilaterals are four sided polygons, with four vertexes, whose total interior angles add up to 360 degrees.

$70° + 120° + 85° = 275°; 360° - 275° = 85°$

25) Answer: C

Factor of 64: (1, 2, 4, 8, 16, 32, 64)

Factor of 36: (1, 2, 3, 4, 6, 9, 12, 18, 36)

Greatest Common Factor is: 4

26) Answer: D

when decimals are compared start with tenths place and then hundredths place, etc. If one decimal has a higher number in the tenths place, then it is larger than a decimal with fewer tenths. If the tenths are equal compare the hundredths, then the thousandths etc.

$0.46 < 0.64$ is correct.

27) Answer: D

Convert decimal to percent by multiply 100: $0.6 \times 100 = 60\%$

28) Answer: A

$4 \times \$11.40 = \$45.60$, and $3 \times \$24.5 = \$73.5$

$\$45.60 + \$73.5 = \$119.10$

29) Answer: A

The rule is $y = 0.40\ x$, when y is output (amount) and x is input (Token).

Amount of 5 token is: $Amount = 0.4 \times 8 = \$2.40$

30) Answer: C

$\frac{7}{9} \times 63 = \frac{7 \times 63}{9} = \frac{7 \times \cancel{63}^{\,7}}{\cancel{9}_{\,1}} = 49$

31) Answer: B

$3 \times 1.4124 = 4.2372$

32) Answer: A

An acute angle is an angle of less than $90°$.

From the options provided, only option A (60 degrees) is an acute angle.

Common Core Math Practice Grade 5

33) Answer: C

One day equals to 24 hours, and 3 days: $3 \times 24 = 72$ hours

$72 \times 1.25 = 90$, $795 - 90 = 705$ liters

34) Answer: A

$58.50 \div 9 = \$6.50$ each roll

35) Answer: B

First, find the missing side of the trapezoid. The perimeter of the trapezoid below is 44.

Therefore, the missing side of the trapezoid (its height) is:

$44 - (8 + 10 + 14) = 44 - 32 = 12$

Area of a trapezoid: $A = \frac{1}{2} h (b1 + b2) = \frac{1}{2} (12)(8+10) = 108$ cm

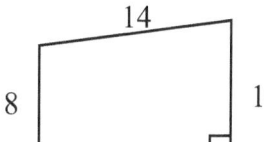

36) Answer: D

Use volume of rectangle prism formula.

$V = length \times width \times height \Rightarrow V = 8 \times 3 \times 4 = 96$ cubic cm

37) Answer: C

There are 10 values on the stem and leaf plot that are less than 40 (8, 12, 14, 17, 19, 20, 21, 21, 27, and 29) and 4 values on the stem and leaf plot that are greater than 75 (84, 86, 89, and 91). Then subtracted 4 from 10, resulting in a difference of 6.

38) Answer: B

To determine each C-value in the table, we should have multiplied 8 by each S-value: ($4 \times 8 = 32$; $8 \times 8 = 64$; $11 \times 8 = 88$; and $14 \times 8 = 112$).

39) Answer: A

Cube is a three-dimensional shape that has equal width, height, and length measurements. If a cube has side length "S" then volume is: $V = S \times S \times = S^3$

$V = S \times S \times S \rightarrow 27 = S \times S \times S$, plug in the small number of options provided to find the length of side. If $S = 3$ $V = 3 \times 3 \times 3 = 27$, means $S = 3$ cubes

40) Answer: C

When shapes are congruent, all corresponding sides and angles are also congruent. The same shape and size, but we can flip, slide or turn. Then, $b = 5cm$

"End"

WWW.MathNotion.com